Tools, Totems, and Totalities

Allen Batteau · Christine Z. Miller

Tools, Totems, and Totalities

The Modern Construction of Hegemonic Technology

Allen Batteau
College of Liberal Arts and Sciences
Wayne State University
Ann Arbor, MI, USA

Christine Z. Miller
Savannah College of Art and Design
Savannah, GA, USA

ISBN 978-981-97-8707-4 ISBN 978-981-97-8708-1 (eBook)
https://doi.org/10.1007/978-981-97-8708-1

This Springer imprint is published by the registered company Springer Nature Singapore Pte Ltd.
The registered company address is: 152 Beach Road, #21-01/04 Gateway East, Singapore 189721, Singapore

Dedicated to my constant companion, Susan Miller, who has supported my professional endeavors for many decades.

To Friends and Family for their support, patience, and encouragement, especially my daughter, Jessica Annelise Knapp, and grandson, Strummer Irvine Vincent Cruz. And to my students: I am always learning from you.

Foreword

Hegemonic orders, as sociocultural formations, are realizations of distinctive constitutive possibilities, especially technological. Certain powers, in the broadest sense, structure—enable and constrain—what patterns of what norms and goals are to prevail, for good or bad. Allen Batteau and Christine Z Miller identify those dominating, defining possibilities or powers, in succinctly synoptic alliteration, as "tools, totems, and totalities." This is wording that captures and promises the exemplary analysis to follow. Theirs is analysis beyond adequate to the challenges, sometimes viewed as existential, they take on.

Those challenges have to do with technology, in its many modern, institutional, and instrumental modalities, and more specifically, with technological risk and cost. They have to do with how best to manage risk and cost—how optimally to select for fitness among technology's "adjacent" or future possibilities, as Stuart Kaufman might put it. They comprise a problem ecosystem perfectly suited for scholars with apposite anthropological sensibilities and sentiments, who also happen to bring backgrounds in engineering (Prof. Batteau) and design (Prof. Miller) to their analyses and advocacies. To be sure, technology's contributions to hominin evolutionary success over the millennia is largely the story of our species' emergence and rise to preeminence. But humans have also exhibited consistently throughout their history a stubborn, precariously naïve confidence—at times an unduly reverential faith—in the infallible efficacy and benignness of their tools, techniques, and technical projects. The burden of this study is to expose just this record in the present period and to suggest remedies as we move precipitously forward, providing indispensable insights and proffering invaluable guidance.

The set of emergent technological problems has various names, reflecting various facets of what confronts us: "Disruption" has been frequently used, at least since the mid-1990s, and while originally denoting innovative if inconvenient change that displaces organizational orthodoxy, it now broadly labels the often inobvious and even insidious ways technology deceives, deflects, deters, disorients, dislocates, distances, disconnects, and otherwise disappoints us in our daily lives and, threateningly, into our realistically imaginable futures. "Polycrisis" is a more recent, fashionable, morphemic fusion intended to convey the sheer diversity and growing urgency of convergent issues (climate change, global conflict, migration, pandemics, etc.) attending our interdependence with tool cultures. "Hegemony" as chosen for this collaborative effort, partly a logical continuation of Prof. Batteau's earlier work on technology's articulation with the commonweal, draws out the integuments and sinews of the digitally mediated systems of control, organizational, and ideological, that govern hearts and minds in contemporary life.

Disruption, the polycrisis, and hegemony are addressed, in their manifestations and implications, with a clarity, sophistication, and thoroughness needed to make a difference to readers, needed perhaps to push them to further inquiry if not action. Timeliness and breadth of focus distinguish this work. The deft, narrative interweaving of relevant theorizations and concepts, comparative sociocultural experience, multidisciplinary diagnoses, and sobering projections—all essential parts of their argument—is balanced with hopeful, salutary calls and counseling for a sustainable coexistence with technology. Primary endorsements include the restoration of truthfulness and trust in productive and distributive institutions, judicious oversight of the rapidly advancing artificial and generative intelligence potentialities, and the commitment to prioritize community values with equitable social sharing.

Technology is not reducible to tools or toolkits, the authors stress, but is properly understood also as the institutional context for their development, management, and use—a normatively ordered environment of power-conditioned relations and practices. Of course, tools are presupposed and figure centrally, as the means and ends of the hegemony they posit. Perhaps oldest of the tools with expansive, open-ended versatility in human evolution is semiosis, or the employment of signs for purposes of representation and communication, for creating meaning. Fifty years ago, Jack Goody advised anthropologists to increase their attention to semiosis, calling systems of communication a "technology of the intellect" and "basic to all social institutions, all normative behavior." Signs, especially symbols, as representational conventions, empower and liberate. They permit the imagination, with few limits, of other times and places, the deliberative selection between alternative actions, the remembrance and avoidance of mistakes, the negotiation of social cooperation and competition both, and the authoring of our own identities and purposes. Vehicles of the transmission of experience, they fix and make public our thoughts and wants, exposing them to reflexive examination. But signs can also disable and dominate, displacing or confounding our understandings and intentions with the self-serving manipulations of others with other, perhaps incompatible, interests. Notoriously, they permit us to dissemble and evade, and to lie.

Life with technology has become life in a totality. Tools, representations in particular, perfuse the whole of our existence and comprehend the content and character of our being. The totems referenced in the title, ever smarter devices, for example—technological objects of distinction, worship, fetish, identification, pride—occupy our consciousness and appropriate our desire, tempting our vanities and our mystification with the allure of faux-fulfillment. The utensils of collective coordination—texts and email, blueprints and maps, instruction manuals and zoom calls, memoranda, metrics, spread sheets, surveys, advertisements, action plans, branding, schedules, annual reports—saturate our subjectivities. Yet like fish only dimly aware of their watery milieu, we tend not to notice the extent of our saturation; much as in an eclipse, the technological totality is a near complete obscuration but for the light shown upon it.

Professors Batteau and Miller compellingly cast that essential light upon the shadowy specter of technological hegemony, connecting the many implications of instrumental dominion to the decline of societal integrity through the effects of spreading misinformation and disinformation, alerting us to their elitist and corporate purveyors, and warning us of their respective dangers for social stability. In Gregory Bateson's felicitous sense of "information" as any difference that makes a difference, the reader will find this book—critical, difference-making anthropology at its best—deeply, importantly informative. It will, as anticipated above—and indeed it should—make a difference to anyone—hopefully to all of us—concerned for our shared prospects.

July 2024

Jordan Pollack
Professor Emeritus
Kyushu University
Ann Arbor, MI, USA

Preface

The dilemma of "technology" in the modern world is that we have a love/hate relationship with technology, both enjoying the advantages it creates but also feeling awkward and uncomfortable with its consequences, whether in the capabilities that technology creates, or the hazards that it also creates. The powers that technology creates for malign actors are a threat. The paradox is that, at the same time, we appear to be incapable of controlling the growing impact technologies have on our "private" lives, our relationships, other species, and on the planet. If we have the occasion to pause to reflect on the consequences—either intentionally or unintentionally—we tend to push away the dread and look instead for another technology to fix what has become a concern (i.e., privacy). In Chap. 11, we write about how continuing to build upon existing tightly coupled technological systems creates a brittleness that leaves legacy systems vulnerable to "normal accidents" (Perrow, 1999) or malicious attacks by bad actors.

When we describe technology as hegemonic[1], we are referring to *socio-technological systems* such as the Internet, electrical systems, smartphones, and automobiles that are widely used and integrated into daily life, shaping cultural values, norms, and practices, and influencing how people think and behave. These systems often facilitate the control and distribution of economic resources, reinforce existing power structures, and limit choice. They can be used to maintain political control and surveillance, influencing governance and policymaking. Finally, these systems become deeply embedded with high barriers to entry that are difficult for competing or alternative technologies to overcome. At the same time, as we note in Chap. 11, they are vulnerable to disruption.

We draw on the work of Antonio Gramsci (1891–1937) who introduced the theory of cultural hegemony. Gramsci, an Italian journalist, social activist, and Marxist who was imprisoned for resisting Mussolini's fascist regime, used the construct of

[1] The concept of hegemony has been applied in various contexts beyond Gramcsi's original use in referring to the power dynamics of developed capitalist systems. A recent book by Alex Williams and Jeremy Gilbert (2022) used hegemony to describe how the combined power of Big Tech and Wall Street "won the world," arguing that "We cannot change anything until we have a better understanding of how power works, who holds it, and why that matters."

cultural hegemony to explain dominance and subordination in developed capitalist societies (Martin, 2002). Gramsci's theory of cultural hegemony describes how a ruling class maintains control through "intellectual and moral leadership" establishing its worldview as the norm, making it appear natural and inevitable. Many scholars have written about and debated Gramsci's ideas. We do not claim to have expertise related to Gramsci's work nor propose to offer any commentary beyond appropriating the concept of hegemony to describe a category of technology that has become increasingly dominant in American society. Technology that falls in this category has evolved over time and exhibits a set of characteristics some of which we described below.

We use the construct of technological hegemony to describe what has occurred through a complex rubric of dynamic conditions and forces that we discuss throughout the book: the Industrial Revolution and emergence of the capitalist system, the rise of standardization and mass production, and cultural narratives that celebrate technological progress and innovation and promote the acceptance and normalization of certain technologies. Finally, globalization allowed for the diffusion of some technologies and socio-technical systems across borders. Each of these factors is addressed in the chapters that follow.

In Chap. 5, we offer an alternative approach to designing technology that we refer to as *convivial*. Building on the original Latin *convivium* meaning "banquet," we describe conviviality as companionable, sociable, and adaptable. In other words, technology that is not only "user-friendly" and flexible in use, but also adaptable and maintainable. This aligns with the shift, though as one of our readers suggest, often at most aspirational, toward human-centered design discussed in Chap. 4. Design has yet to reach the promise of true "user-centeredness."

This book has been a collaboration and conversation between two coauthors from different backgrounds, generations, and life experiences. We share perspectives because of our previous relationship as professor and dissertation chair (Allen Batteau) and Ph.D. student (Christine Miller). We have arrived at common ground from different places. If at times we are lacking a single clearly aligned voice that is the consequence of the choices we have made in thinking through this project and ultimately committing it to writing.

Ann Arbor, MI, USA — Allen Batteau
Savannah, GA, USA — Christine Z. Miller

References

Forgacs, D. (Ed.). (2000). *The Antonio Gramsci reader: Selected writings 1916–1935*. New York University Press.

Gramsci, A. (1971). *Selections from the prison notebooks* (Q. Hoare & G. N. Smith, Eds. & Trans.). International Publishers.

Martin, J. (Ed.). (2002). *Antonio Gramsci: Critical assessments of leading political philosophers*. Routledge.

Williams, A., & Gilbert, J. (2022) *Hegemony now: How big tech and wall street won the world (and how we win it back)*. Verso.

Acknowledgements

Writing a book is never a solitary endeavor. Our efforts on *Tools, Totems, and Technology* have benefited from the comments and criticisms that our colleagues have generously shared with us. At the top of the list would be Bradley Trainor, whose comments as we moved along were a constant source of guidance and inspiration. Also to be mentioned would be Jordan Pollack, an anthropologist with considerable experience in Japan, and Bill Beeman whose ethnographic experience in contemporary societies has contributed much to our endeavors, and who have contributed forewords to the volume. Julia Gluesing's insights into technological development have contributed much, as did the comments of Carolyn Psenka and Nick Schroeder. We express our greatest appreciation to these colleagues, who deserve credit for the success of our book. Its inevitable shortcomings remain our responsibility.

Ann Arbor, MI, USA — Allen Batteau
Savannah, GA, USA — Christine Z. Miller

Contents

List of Figures

List of Tables

Chapter 1
Introduction: Technology and the Modern Imagination

Abstract In this book, we interrogate that imagination and the ways we have built the imagination of technology into the world around ourselves. This book represents an ethnographic expedition in search of the core idea, perhaps the key symbol, that permits us to designate such diverse objects as video games, X-ray machines, thermonuclear weapons, space probes, electric power grids, and monumental architecture with the common denominator of "technology." We explore and critique the near-universal faith in "progress" that derives from modernist assumptions and thereby afford these objects with a common set of attitudes and a belief that they represent progress. These attitudes include respect for their usefulness, awe at their great powers, mystery in their functionality for most of the population, and certainly an expectation that these devices will make our lives more comfortable, more secure, more enjoyable, and more fulfilled. When our devices fall short of these expectations, it simply means that we must redouble our efforts: improve the design, install a more powerful microprocessor, write some new lines of code, or fine-tune the "human factor."

Technology's contribution to human progress is mostly imaginary. The wondrous technological devices that multiply productivity, that create social connections spanning oceans and leaping over mountains and plains, and that enlarge the mind to encompass vast quantities of information represent an advancement in the human condition only to the extent that those who make or use them can *imagine* a perfection of productivity, connectivity, and infinite knowledge—a god-like transcendence of the mortal human condition. If such an imagination is called into question, then the axiomatic equation of "technology" and "progress" collapses.

This equation, of course, has been a cornerstone of Western thought for nearly two centuries, and this perfection has been sought for millennia. Although the happy marriage of empirical philosophy and the useful arts can be traced back to Francis Bacon, it was not given a name until 1823, nor broadly applauded by the public before the twentieth century. In the eighteenth century, "progress" was primarily a musing of *philosophes*, and the industrialization of the useful arts horrified broad segments of the population. Large and substantial groups in the society—aristocratic gentlemen, Lake

A. Batteau and C. Z. Miller, *Tools, Totems, and Totalities*,
https://doi.org/10.1007/978-981-97-8708-1_1

District poets, Manchester millhands—saw what would later be called "technology" as a deterioration of civilization, a reversion to barbarism. The enshrinement of technology as the salvation of the human condition, a remedy for illness, mortality, hunger, and discomfort, required an attitude that trusted efficiency, privileged the individual, and distrusted society—in short, a modern sensibility. Technology is a creature of the modern imagination.

Likewise, the near-universal faith in "progress" derives from modernist assumptions. Anthropologists point to the undeniable facts of cultural evolution over the past hundred thousand years, a drive toward increasing differentiation, increasing sophistication of tools, increasing use of external sources of energy, and increasing domination of the many by the few. Whether such modern marvels as alienation, totalitarianism, environmental degradation, or mass warfare also should be considered part of the sunlit uplands of progress, is a question for moral philosophers to debate. Considerable imaginative effort and amnesia is required to see the twentieth century as an unambiguous improvement over those that came before.

The imagination of progress is a recent phenomenon. For most of human civilization, the arc of history has been either cyclical or downward; only with the enlightenment of the eighteenth century did Europeans imagine an upward progression of history. Part of this upward progression was the Industrial Revolution.

In this book we interrogate that imagination and the manner in which we have built the imagination of technology into the world around ourselves. This book represents an ethnographic expedition in search of the core idea, perhaps the key symbol, that permits us to designate such diverse objects as video games, X-ray machines, thermonuclear weapons, space probes, electric power grids, and monumental architecture with the common denominator of "technology" and thereby afford them a common set of attitudes and an assumption that they represent progress. These attitudes include respect for their usefulness, awe at their great powers, mystery in their functionality for the majority of the populace, and certainly an expectation that these devices will make our lives more comfortable, more secure, more enjoyable, and more fulfilled. When our devices fall short of these expectations, it simply means that we must redouble our efforts: improve the design, install a more powerful microprocessor, write some new lines of code, or fine-tune the "human factor."

These attitudes of awe and mystery and expectations of extreme functionality set "technology" apart from mere tools and other useful artifacts. In our lives there are many useful objects that we seldom think of as "technology." These include eating utensils, gardening tools, lawn chairs, and the bric-a-brac with which we fill our leisure time. Earlier societies had tools and useful artifacts, some of which were quite ingeniously devised. Anthropologists sometimes characterize these as "primitive technology," although in so doing they superimpose a modernist term and set of attitudes on ancient objects and societies. "Technology," as we will make clear in this book, has a set of attitudes and expectations *not* associated with primitive tools. The creation of technology lies in the manner in which modern society has selected and refined certain useful tools and objects and has invested these tools and objects with both a broad array of functionality and an equally broad array of expectations including magical powers. This investment of functionality and expectations has created for us a set of anomalies and conundrums that we wish to explore as part of our ethnographic expedition.

The anomalies of technology include not only the "productivity paradox" that received considerable discussion in the 1990s and which we discuss in Chap. 8, but also the paradox of "labor-saving technologies" that actually create more work. They also include monumental, hypertrophic technologies that demand investments of billions of dollars for the achievement of goals that are irreducibly non-technological. One example is the "virtual border wall," a Maginot Line[1] in low-Earth orbit using satellites and sensors and electronic surveillance, intended to keep out illegal immigrants, most of whom have simply overstayed their visas. It also includes technological arms races, such as the "race between hardware and software" or the "race between bandwidth and content" (Batteau, 2001), the fact that as the Internet is used for more and more applications, the *availability* of the Internet in all corners of the globe comes into question. It also includes techno-narcissism, whether displayed in fast cars, powerful computers, or multifunction cell phones, the achievement of any of which has as much to do with personal display and expression as with a more just society or a more prosperous world.

Any of the devices we collect can be used to fulfill multiple social objectives, and the choice of using high-powered "smart" sensors to exclude illegal aliens rather than to monitor global warming, or using terabyte computer memory to accumulate intimate personal detail rather than to crack viral genomic codes, or to use internal combustion technology to overpower large SUVs rather than to propel fuel-efficient transportation appliances is a social choice, not one that is inherent in the device. Yet such choices ramify and replicate themselves, so that for some commentators we are confronted with technology out of control (Winner, 1995), turning on its creators like Frankenstein.

This is an ignorant perception. It is ignorant of the nature of the devices and of their internal mechanisms, and it is ignorant of the dynamics of complex adaptation and institutional ecosystems that set them on their "out-of-control" courses. Most importantly for our present purposes, it is ignorant of the manner in which modern society has created technology in its own image.

Words matter. The fact that a specialist term—"technology"—has passed into common usage indicates a profound cultural change. Having a common term for such devices and activities as computers, microwave ovens, internal combustion engines, televisions, and genetic manipulation indicates a set of common qualities or attributes for these devices and activities, if the term has any integrity. It also *suggests* a common set of attitudes and expectations—a common mood, if one wishes for these devices. These attributes and attitudes are not always readily apparent.

Perhaps the most commonplace assumption about technology is *instrumentality*—the assumption that, of course, it is useful, or intended to be so. Utility in the popular view defines technology. However, this is a slippery definition. Not everything that

[1] The Maginot Line was a notorious line of fortifications along the French-German border, built in the 1930s. Intended to keep Germany out of France, it became a symbol of technological failure when the German army bypassed it and invaded France through Belgium.

is useful—rags, eating utensils, digging sticks—is considered technological, and not everything that is defined as technology is useful or intended to be so—millions of lines of object code lie around lacking employment because they do nothing useful. Are we to banish such software from the realm of technology? Undoubtedly, at some point some programmer *intended* that software would be useful, but to depend a definition of "technology" on the intentions of its creators means that, potentially, any object can be seen as "technology," and the usefulness of the term collapses into a heap of subjectivity.

It is similarly the case with the understanding of efficiency and productivity. From the opposite perspectives of critical theory and managerial ideology, diverse thinkers such as Andrew Feenberg (2014) and Alan Greenspan (2000) arrive at a similar conclusion: that technology's utility—supposedly its core value—can be comprehended only in a social context, that is only in conjunction with other, *non-technological* values.

For nearly a century, the axiomatic equation of technology and productivity, and "productivity" and "progress," has been used to justify massive investments in technology. In 2019 the federal government spent more than $68 billion on research and development,[2] private industry spent nearly $498 billion, and nonprofit organizations another $25 billion. Nearly 8% of the federal budget of $4.4 trillion[3] or 3% of GDP of approximately $21.43 trillion.[4] Rarely was any of this motivated by any curiosity concerning the fundamental laws of nature.

If technology per se does not produce productivity improvements, and if many productivity improvements can be achieved using less technologically intensive approaches, then why do we massively spend and invest in technology? What are we buying? If not productivity, then what are the values that technology delivers?

This is the core question of this book, and the answer that we sketch here will be developed in full color in the next thirteen chapters.

Like so many other goods and services in a modern economy, technology supplies a need that it also creates. "Invention is the mother of necessity," as the saying goes. Ever since the enlightenment, when supernatural explanations and priestly powers were overthrown, we have lived in a world in which mystery was banished to the sidelines, a world where the sources of awe and reverence are suspect. In popular entertainment, we have bionic androids and mythic heroes with awesome powers, but almost no one takes these portrayals as serious descriptions about how the world works. We have public leaders who believe in faith-healing and the facticity of the book of Genesis, but these assertions are tolerated with polite smiles, not accepted as the conceptual foundations of medical science or history curricula. We have had financial alchemists, who spun derivative gold out of securitized portfolios, but after October 2008, when their computerized models and cryptocurrencies collapsed, their

[2] U.S. R&D expenditures, by performing sector and source of funding: 2010–22 for federal, private sector, and nonprofits (https://ncses.nsf.gov/pubs/nsf24317).

[3] Office of Management and Budget (OMB.gov) https://www.govinfo.gov/content/pkg/BUDGET-2019-PER/pdf/BUDGET-2019-PER.pdf

[4] World Bank Group (https://data.worldbank.org/indicator/NY.GDP.MKTP.CD?locations=US).

credibility is on par with that of astrologers. Society—the connections among the millions of souls who make up a community or a nation—has lost its magic.

Into this void steps technology. Technology supplies the magic that is missing in our society. The fact that anyone can climb into the clouds, or hear a distant friend, or, like the sorcerer's apprentice, produce thousands of identical objects is just magic. In society at large, where only a select few understand the mysteries of aerodynamics, diversified portfolios, semiconductors, or supply chain management, these capabilities are accepted as modern marvels.

It is simple to document that magical thinking is associated with technology. Most times this is by an ignorant public and politicians, who assume that such complex and energy intensive devices as a space shuttle or a missile shield or an econometric model will be cheap to produce and easy and reliable to operate. At times it is by charlatans who prey on this eagerness to be duped. Thus, we have a billion dollar electronic wall along a few hundred miles of the thousand-mile border with Mexico in order to keep out aliens (most of whom enter the country legally), or a multibillion-dollar "ballistic missile shield." The assertions that these technologies might keep America safe can most politely be described as magical thinking. This enchantment with technology is a testament to its hold on contemporary society.

At this point the educated reader, perhaps with an engineering degree, will scoff and say "I may not be able to explain everything about Bernoulli effects or von Karman vortices, but I know experts who can, and they can assure you that there is nothing magical about flight. It is all very scientific." For this we can reassure our educated reader that with this understanding he is placing himself in the company of a learned elite, thus demonstrating the second value that technology affords: a democratized mastery. In modest measure for the masses, and in ample supply for a small elite, technology creates mastery over space, over time, over fellow citizens, and over the laws of nature, and thus banishing, however, locally and momentarily, the overwhelming sense of insignificance communicated by a mass society.

This sense of mastery is at the heart of techno-fetishism, the idea that technology affords me powers I would not otherwise have. Similar to commodity fetishism in Marx's analysis, techno-fetishism which we discuss in later chapters confounds a relationship with people with a relationship with objects, the objects in question not being so much economic commodities as highly engineered objects compressing vast amounts of energy, information, and functionality.

Technology gives us power and knowledge, however, only insofar as the power and knowledge enter into social circulation, that is, only insofar as they alter relationships with fellow members of one's society. Our ability to climb into the clouds counts for nothing if we cannot use this ability to reconnect, to redefine, or maybe just reimpress our fellows. Our command of a 250 hp 8-cylinder overhead cam engine is pointless unless it is installed in a car that allows one to go faster and farther and in greater comfort than my friends, or better yet to share the experience with them. Our entire ability to transport ourselves in fact is of value only insofar as it connects us with others, either in distant places, or upon our return. One's state-of-the-art camera-videorecorder-PDA-cellphone has no value unless it alters our personal and professional relationships, preferably for the better. Assuming that an airplane, a car, a computer, or any other technology has any value or power or functionality independent of the social and institutional context in which it is used, stored, exchanged, or displayed is the heart of techno-fetishism. Yes, technologies such as airplanes and automobiles allow me to travel farther, meet more people, and see more friends, but the *value* of that mastery comes from the meaning that *not I, but my friends, associates, and others have invested in it*. Just as Marx saw commodities as repositories of social labor, we must see technologies as repositories of social relationships.

The apex of techno-fetishism is in those technologies that allow us to slip in imagination (if not quite yet in fact) the bonds of society. Thus, we have proposals for manned spaceflight to Mars, which will require (at current estimates) expenditures of $230 billion over 20 years, yielding a wealth of scientific knowledge but no commercial opportunities for at least a century. Like proposals for colonization of the Moon, since abandoned, the Mars Mission rests on an assumption that social interest and support can be sustained over several decades of business cycles, changes of administration, collapsing foreign policy, collapsing financial structures, and a notoriously short public attention span. Technologically, travel to Mars, or colonizing the moon, or anything else, is possible provided one has sufficient budget, sufficient schedule, and sufficient willingness to distort other social priorities. The constraints on technological possibility are social constraints, the surly bonds of society, which always have other uses for our time, resources, and attention. The awesome technological achievements of the twentieth century, the atom bomb and the Apollo program, were possible only with the national support afforded by wartime mobilization.

Imagining a society in which technologies' benefits are weighed against their costs in undemocratic concentrations of power, their costs in the neglect of social priorities, and their costs in increased levels of hazard and risk, will require a rediscovery of the social, a reimagination of the magic and wonder that come from connecting not with objects and with the forces of nature, but with fellow men and women.

To begin this exploration, we will first examine the prehistory of technology—or more accurately, the history of toolmaking by earlier civilizations, when *techné* and *logos* existed in separate social universes. Chapter Two considers the prehistory of technology, examining how *techné* was represented at a time when most technicians were illiterate craftsmen. Prior to the Industrial Revolution, technology had no privileged place in history and in fact was subordinated to the subsistence of peasants and the wars of kings.

Humans have been making and improving tools for millennia, but a *science* of toolmaking, as contrasted to the rough-and-ready empiricism of illiterate craftsmen, would have to wait until the nineteenth century. In earlier centuries those who made and used and improved tools, and those who pondered the laws of nature belonged to separate social classes, indeed separate callings, rarely exchanging knowledge. Master craftsmen had their own lore, carefully guarded, demonstrated in the "mystery plays" of the guilds and passed down through apprentices and journeymen. As an account of accumulating practical knowledge, gathering force with the freeing of the mind in the enlightenment and the freeing of the body in the Age of Discovery and the freeing of desires with the Industrial Revolution, this history created a need for a New Science, which Jacob Bigelow (1829) named "technology."

Between the publication of Bigelow's *Elements of Technology* in 1829 and the rapid growth of corporate research laboratories a century later lay the conquest of a continent, the construction of large-scale systems, and associated business empires providing transportation, communication, and electrification to the far corners of America. Prefigured with the rail networks of the nineteenth century, the twentieth century uniquely saw the construction of power grids, universal telephone service, air transport networks, industrial supply chains, and the far-flung business interests of an integrated global industrial economy. In contrast to earlier business empires such as the Hudson Bay Company or the Dutch East India Company, these tightly coupled networks lacked royal charters or state sanction, instead using their command of new symbols—"technology" and "progress"—to legitimate their operations. That these two terms, oddities in the eighteenth century and problems in the nineteenth, are today at the beginning of the twenty-first century axiomatic and demonstrates how successful their effort was. Chapter Three, "The Constitution of Technology," explores the relationship between large-scale, tightly coupled operations and the core elements of the idea of technology. In Chapter Three, we examine the constitution of "technology" in the seventeenth century. According to the *Oxford English Dictionary*, the first recorded usage of the word was in 1621; prior to this there was a discourse of *techne*, but it was not joined to *logos*, the authority of written language. Architects such as Vitruvius (*De Aqueductae Romanae*) wrote about classical structures, but it was not considered "technology."

By the first decades of the twentieth century, technological society was in full flower, and serious thinkers such as Thorstein Veblen could imagine turning society's affairs over to a "soviet of technicians." Yet, as we demonstrate in Chapter Four, "An Engineer's Perspective," the engineering profession has been spectacularly successful because it did not try to solve all of society's problems: it could create impressive transportation and communication devices because it did not pause to consider how transportation and communication can engender new forms of conflict. Ignoring this, it created weapons of unprecedented power, making this a less peaceful world. It created an industrial society of unparalleled abundance, leaving it to others to attend to the hazards that abundance creates. Chapter Four, "An Engineer's Perspective," establishes the boundaries of an engineering approach to technology. Chapter Four considers technology from an engineering perspective, stressing the importance of standards and other forms of autonomous representations for defining technology. *Autonomous representations* are descriptions of the technology that are independent of the actual tool or object or implement yet are at the core of technology as it is understood by engineers.

For nearly a half-century historians, economists, and engineers have explored the relationship between technology and culture, understanding culture as any human creation outside the engineered domain. In these discussions, culture is a residual category, anything that is not explained by the laws of nature. For a specialist in the science of culture, this is no less disturbing than would be an art historian's explanation of photonic junctions: an account that might create some amusement, but certainly would never be relied upon for engineering development. Likewise, technologists' statements regarding the social benefits of technology should carry no more authority than art historians' opinions of the production of Mandelbrot fractals. Both are undoubtedly interesting but fundamentally unreliable.

As an alternative to such anemic understandings of culture by nonspecialists, we intend to supply here a robust understanding of culture, specifically a robust understanding of the culture of technology. In the anthropologist's account technology is one among multiple cultural practices, with its own unique set of characteristics, of which instrumentality is but one of many. As a cultural anthropologist and a design anthropologist, one of whom pursued a career as an engineer for ten years and the other a career as a design educator, we articulate both the practical and the symbolic meanings embedded within any technological device. Some of these are undoubtedly positive, but others such as the degradation of communication are more problematic: communication that relies solely on technological devices undoubtedly has greater reach than face-to-face communication, but the inevitable trade-off of reach and richness (Evans & Wurster, 1999) is rarely commented on.

Chapters Five and Six examine technology from a design perspective, noting how the aesthetic dimensions of technology, notably modernism, figure into popular understandings of technology. For most of history, designers and artists occupied separate social and semantic spaces from rude mechanicals and other artisans. With the advent of modernism in the nineteenth and twentieth centuries, design became a critical element of technology and a critical *instrument* in shaping society and its built environment. Chapter Six introduces influential design thought leaders, some trained as professional designers and others from various disciplines, that suggest the contours and characteristics of design when it is not in the service of industry and commercial interests. It also considers the nature of design in the realms of experimentation and speculative imagination.

Chapter Seven, "The Narratives of the Machine," develops a narrative account of technology, building on techno-celebrants such as Lewis Mumford (1971), techno-skeptics such as Jacques Ellul (1964), and critical technology theory from Martin Heidegger, Jurgen Habermas, and Andrew Feenberg (2014). It points up the fundamentally *irrational* attachment to technology that is resolved by seeing technology as a *symbolic,* rather than a rational construction.

Chapter Eight, "Who Are We?" is an examination of techno-totemism and other aspects of how modern society define their identity around technologies. Techno-totemism is the definition not only of one's identity but of a society's entire cosmological outlook in terms of technological objects. In America, the perfect example of totemic objects is the car, yet many other examples (guns, computers, smartphones, airplanes, and boats) can be found as well. Automobiles not only define *who we are*, but the entire structure of our lives and the built environment. For many Americans, *guns* are valued less because they make families safer (which they demonstrably do not) but rather because they are tokens of identity, of who we are as Americans (See Chapter 3 for an image of a contemporary fetish in Fig. 3.4) (Slotkin, 1992).

Chapter Nine examines the relationship between technology and economic values, focusing on the "productivity paradox," the undisputed fact that many "labor-saving" technologies create more work, at least for the women of the society. The "productivity paradox" results from a focus on private goods at the expense of public goods and common pool resources, which has been the focus of liberal economics for the past three centuries. By considering other types of values, including both sociability and transcendence, the value added by technology becomes more subtle.

The relationship between technology and citizenship is considered in Chapter Ten. "Things are in the saddle and ride mankind," Ralph Waldo Emerson wrote in 1847, at the dawn of our technological era. By taking a critical rather than enthusiastic view of technology, we can begin to see how technology makes us all slaves of our own devices.

Chapter Eleven, "The Emperors' New Clothes," enlarges the aperture to examine how technology today shapes the human community. Many technologies today—space travel, social media, cryptocurrency—promise effortless gratification and abundance. A major innovation of the twentieth century was the creation of large-scale technologies, spanning not simply continents but the entire globe. Yet these technologies also create new peripheries, leading to greater global instability. Examples of large-scale technologies include air transport, manufacturing supply chains, and, of course, the Internet.

"A More Brittle World," Chapter Twelve, draws on extensive research into industrial disasters. As Charles Perrow demonstrated in Normal Accidents (1999), when a system is complex and tightly coupled, accidents can be expected and hence are "normal." Many users of technology, particularly in high-risk industries such as air transport, understand this, and frequently opt for "low-tech" solutions. This chapter examines how we have created a more brittle environment over the past fifty years, perhaps most notably in complex supply chains spanning the entire globe.

Chapter Thirteen, "The Enchantment of Technology," examines the cultural roots of our love affair with technology. Drawing on Marshall Berman's insights in *All that is Solid Melts into Air* (1988) we examine the close relationships among technology, modernity, and modernism. Technology, we suggest, is as much about *images* as it is about functionality, and the chief image or style associated with technology is modernism, an artistic movement that came into being only in the early twentieth century. The "Metaverse"—Facebook's construction of its new virtual reality—is the perfect representation of this love affair with technology.

In Our conclusion, "Where We Go from Here" suggests how we might imagine a world no longer in thrall to technology. When terrorists armed with box-cutters can bring the most technologically sophisticated nation to its knees, then we have proven the limits of technology. America's reaction to September 11, 2001, suggests a brittle prosperity. By viewing technology not as an autonomous force, but as simply one of multiple strategic options for a society, one whose limits have been demonstrated, we can begin discussions of a society, one that no longer presumes that there is a technological solution for every ill. Our objective with this book is to stimulate a dialogue regarding the alternatives to the modern construction of hegemonic technology as a richer palette for the solution to the problems of contemporary society, problems which include inequality, resource exhaustion, and the concentrations of power in wealthier societies.

In 1973 Daniel Bell, in *The Coming of Post-Industrial Society*, predicted that over the next 50 years the economy of goods and services would be replaced by an economy of information and connections. Today technological empires such as Google and Facebook traffic not in goods and services but in information, and Facebook has monetized "friend"-ship for sale to advertisers. How the algorithms of Facebook, Twitter, Amazon, and many other systems manipulate user behavior, promoting products and corrupting our politics, will become one of the central dramas of tomorrow. Just as Bell anticipated a society beyond industrial society, so too now we open the door for a society beyond technological society, one in which human connections count for more than sophisticated tools and instrumentalities. Imagining a society not enchanted by its tools and instrumentalities is, we suggest, one of the leading questions of our time. Just as the Neolithic Revolution of 14,000 years ago moved humanity beyond foraging toward agriculture, and the Industrial Revolution moved humanity into new forms of subsistence, and in Daniel Bell's perspective in *The Coming of Post-Industrial Society* (1973) moved humanity from the production of goods and services into the production of information and images, so too reimagining technology in service to humanity and the planet will move us away from production to connection, toward a rediscovery of the social, in which connections count for more than efficiency, and transcendence counts for more than usefulness. The (non-Newtonian) dynamics of transcendence, where inspiration and enlightenment have far greater value than material goods, will be the reward of a society beyond the modern construction of hegemonic technology.

With these data points in place at the start of our expedition, we move to Chapter Two in which we consider the prehistory of technology, examining how *techné* was represented at a time when most technicians were illiterate craftsmen. We explore how prior to the Industrial Revolution, technology had no privileged place in history and in fact was subordinated to the subsistence of peasants and the wars of kings.

References

Batteau, A. (2001). The anthropology of aviation and flight safety. *Human Organization, 60*(3), 201–211.

Bell, D. (1973). *The coming of post-industrial society: A venture in social forecasting*. Basic Books.

Bigelow, J. (1829). *Elements of technology*. Hilliard, Gray, Little, and Wilkins.

Emerson, R. W. (1847). Ode, inscribed to William H. Channing. In ***Poems*** (pp. 166–169). Phillips, Sampson and Company.

Evans, P., & Wurster, T. S. (1999). *Blown to bits: How the new economics of information transforms strategy*. Harvard Business School Press.

Feenberg, A. (2014). The ruthless critique of everything existing: Nature and revolution in Marcuse's philosophy of praxis. In R. Aguirre (Ed.), *The Bloomsbury companion to Herbert Marcuse* (pp. 127–144). Bloomsbury Academic.

Greenspan, A. (2000, July 11). Structural change in the new economy. Speech presented at the U.S. Department of Agriculture's Agricultural Outlook Forum, Washington, D.C. Retrieved from https://www.federalreserve.gov/boarddocs/speeches/2000/20000711.htm

Mumford, L. (1971). *Myth of the machine: Technics and human development*. Mariner Books.

Perrow, C. (1999). *Normal accidents: Living with high-risk technologies*. Princeton University Press.

Slotkin, R. (1992). *Gunfighter nation: The myth of the frontier in twentieth-century America*. University of Oklahoma Press.

Winner, L. (1995). Do artifacts have politics? In D. MacKenzie & J. Wajcman (Eds.), *The social shaping of technology* (pp. 26–38). Open University Press.

Chapter 2
The Prehistory of "Technology"

Abstract In this chapter, we examine the engineering accomplishments of Western society before the concept of "technology" appeared. These accomplishments, some quite impressive and enduring even today, suggest a historical specificity for "technology," an invented concept from the nineteenth century as more than simply an aggregation of tools and instrumentalities. We posit that technologies always exist in an institutional context, just as the concept of "technology" exists in a specific historical context. As obvious as this is, many efforts to deny or at least ignore the institutional context of technologies abound. We show that the building blocks of technology, like the bricks and beams in the construction of a cathedral, were all present at the outset of the Industrial Revolution making possible the assembly of the edifice that we now label as technology, a word that was only coined in the seventeenth century.

Technologies always exist in an institutional context, just as the concept of "technology" exists in a specific historical context. As obvious as this is, many efforts to deny or at least ignore the institutional context of technologies abound. Space travel, for example, requires not just rockets and space suits, but also a vast institutional infrastructure of research, training, public acclamation, and funding, sustained over decades if not centuries, and not simply the starry-eyed dreams of the Elon Musks of the world. Institutional infrastructures are inevitably *social* formations, requiring participation, legitimation, compromise, and adjustment of thousands if not millions of individuals. Power grids, a common pool resource (i.e., an institution) assuring electric power to millions of households, require cooperation among multiple utilities and municipalities, and when this cooperation fails, as it did in 2021 when Texas declined to join the Western Interconnection, a power grid encompassing eleven states and two Canadian provinces, from British Columbia to New Mexico, power fails. Further, these two, the technologies and their institutional ecosystem, coevolve; as one becomes more complex, the other must do so also, or else collapse into a heap of irrelevance or dysfunctionality.

In this chapter we examine the engineering accomplishments of Western society before the concept of "technology" appeared. These accomplishments, some quite

A. Batteau and C. Z. Miller, *Tools, Totems, and Totalities*,
https://doi.org/10.1007/978-981-97-8708-1_2

impressive and enduring even today, suggest a historical specificity for "technology," an invented concept from the nineteenth century as more than simply an aggregation of tools and instrumentalities.

Institutions constitute the durable peace treaties among multiple human communities. Primitive hunting bands did not have *institutions* so much as they had customary ways of life; it was only when these hunting bands were aggregated into tribes and nations that they found the need to develop institutions, embodying authority, settled rules, and a sense of "who we are," as Mary Douglas describes in *How Institutions Think* (1986). In modern society, the institutional environment is no less important to survival than the natural environment.

The different constituents of institutions—authority, roles, and identities—are intimately linked and require adjustment to function. Automobiles are a perfect example, and efforts to extend them beyond their existing institutional ecosystem—flying cars, for example, or self-driving cars—have been unsuccessful to date, not because of their technological limitations, but because of their institutional maladaptation. When technologies push the boundaries of their institutional environment, controversy is inevitable. Contemporary debates over social media and privacy are perhaps the best current example.

Examples of this failed coevolution of technologies and institutions abound: Railways, for example, originally were used for transport from mine mouths to rivers. The use of railways on a continental scale corresponded first to a scaling up of industrial production from local factors to continental supply chains, plus the expansion of the national writ to encompass the entire continent. Railways strangled local communities, as Frank Norris described in his novel *The Octopus*, and eventually provoked the federal government to create the Interstate Commerce Commission in 1887 which regulated freight rates and rail safety. Similarly, the Wright Brothers' invention of the airplane led in a few years to the creation of the Federal Aviation Administration in 1926 and eventually vast national and international institutions. Flight safety, an issue that only concerned two brothers in 1903, has become an international issue as air travel knits together the entire world, with organizations such as the Federal Aviation Administration and the International Civil Aviation Organization providing guidance and regulation around the world. Air travel, the safest form of travel available (in terms of fatalities per passenger mile), is more hazardous the further one gets from core nations, due mainly to the institutional environment on the periphery.

Institutions are the social edifices that enable members of a society to live together. They encompass authority, division of social roles, but also a sense of *who we are* (Douglas, 1986). Institutions are the building blocks of societies no less than roads and bridges, and to understand where our society goes beyond the hegemonic technology, we need to understand its institutional infrastructure.

The causal linkage works in the other direction as well. The scaling up of the writ of the federal government, during the Great Depression and World War II and subsequently the Cold War led to, or at least accelerated, the development of aircraft and alternative fuels and nuclear power, and the space race between the United States and the USSR led to the first satellite, the sputnik, and eventually the landing on the moon.

In this chapter, we examine how tools evolved before there was the institutional edifice that today we call "technology." Far from being a natural or an inevitable evolution, the emergence of "technology" and the corresponding institutionalization of its components was perhaps a historical accident, occasioned by several events or trends in the fifteenth, sixteenth, and seventeenth centuries. These events included a growing body of engineering and scientific writing as exemplified by Copernicus and da Vinci, the corruption and decline of the feudal system, Europe's discovery of a New World, and the enclosure movement of the seventeenth and eighteenth in which illiterate peasants were displaced from the commons and into the emerging factory towns of Manchester and Birmingham, new domestic arrangements previously associated with armies on the march. These resulted in both the Industrial Revolution in the eighteenth century and the emergence of a new term and concept, "technology." This term and concept united the authority of the written word (*logos*) with the skill of the (most typically illiterate) craftsman, *techné*, and created a dynamism that has turned the world upside down.

Some of the most important authors bringing together *techné* and *logos* include Marcus Vitruvius Pollio, whose *The Ten Books of Architecture* (2020) established some of the basics of engineering practice both in the classical period and into the Middle Ages. A review of its table of contents will give a picture of its wide-ranging interests.[1] Similarly Frontinus (2004) discusses the network of aqueducts bringing water to Rome from several hundred miles away (the South of Gaul, for example) and both engineering (design) and management issues (water allocation). The Roman aqueducts might be considered a precursor to the power grids of two millennia later (Bakke 2016) and in fact posed similar management issues.

These authors established a rich legacy of writing on tools and engineering accomplishments that would fill an entire library. For example, Vitruvius covered such diverse topics as city sites, walls, streets, and sites for public buildings. He covered building materials, building design, public buildings, private dwellings, and engines for raising water, among many other topics in several hundred pages. Vitruvius is probably best remembered for his portrayal of "Vitruvian Man," pictured here, which displayed the idealized proportions of the human body. For Vitruvius, and many since, the human body provided an idealized edifice, a model for engineering (Fig. 2.1).

All of this is well established in the literature of the history of engineering and in fact would fill this volume several times over to cover adequately. What is *not* well established is the missing link between *techné* (making or doing, craftsmanship) and *logos* (the authority of the written word). This link was needed to make engineering a respectable discipline worthy of being taught in academies and later universities. Although the first academic institutions date back to Plato's Lyceum in the fourth century BCE, and the first universities beginning with the university in Bologna in 1088, it was not until the nineteenth century that *Institutes* of technology and in fact

---

[1] To give a brief overview of Book III (out of 10), its chapters include On Symmetry: in Temples and the Human Body, Classification of Temples, The Proportions of Intercolumnations and of Columns, the Foundations and Superstructures of Temples, and Proportions of the Base, Capitals, and Entablature in the Ionic Order.

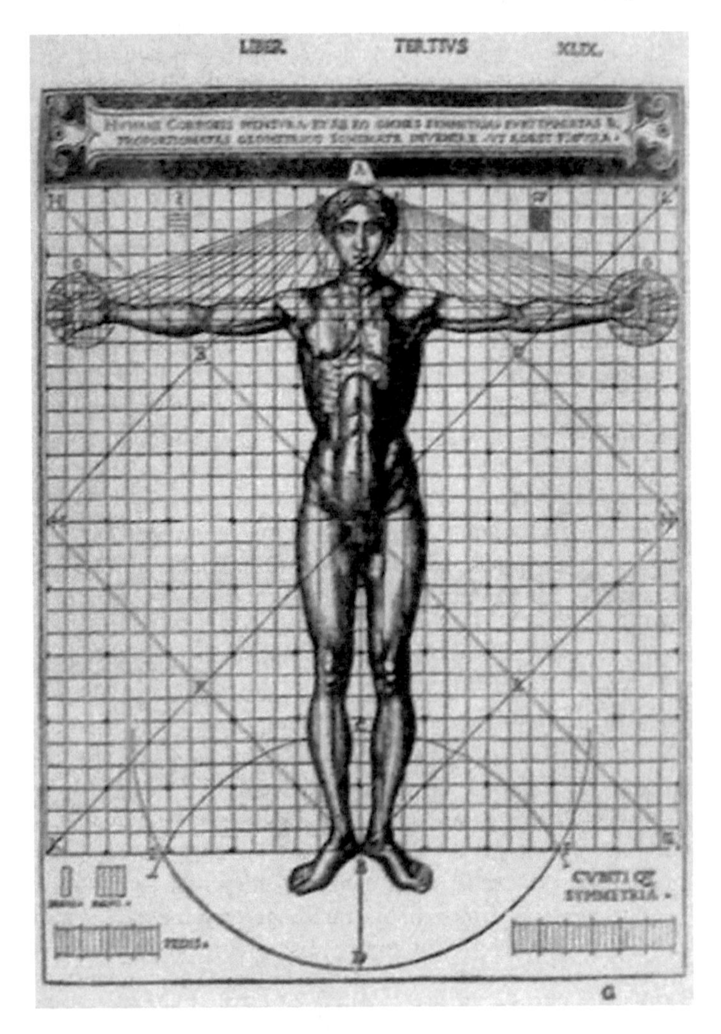

Fig. 2.1 Vitruvian Man, Leonardo Da Vinci

the entire coinage of the word "technology" came along. Institutes of technology, beginning with the Massachusetts Institute of Technology (founded in 1861) and later with technological institutes and universities around the world propelled a new dynamism for society.

The story we seek to tell here is less of the accomplishments of *techné* over the ages and more of the *institutionalization* of these accomplishments, first in craft guilds and later in government regulation, and finally in the establishment of the concept of "technology," whose constitution is more fully described in our next chapter. As everyone knows, technology has transformed the world, yet "technology" did not exist before the modern era. Before Jacob Bigelow's *Principles of Technology*

(1829), humanity was a tool-using animal, but *techné* was disconnected from the authority of the written word.[2]

Characterizing a toolkit as "technology" creates new spaces for exploration and exploitation: it opens the door not only to illiterate craftsmen but also to learned scientists who explore the laws of nature. "Technology" in modern society is an institutional edifice that did not exist in earlier societies. Combining an understanding of the laws of nature with the practicalities of toolmaking, whether those tools be internal combustion engines in transportation devices or search engines on the Internet, has created a dynamism both for transportation and for exploration of the natural world.

Building an Engineered Edifice: Roads, Bridges, Monuments

Part of this institutionalization is the recording of engineering methods and accomplishments in written records, books, and libraries. A library is more than simply a warehouse for books. A library is a monument to learning over the ages, and libraries are far older institutions than many of the Roman engineering accomplishments. For example, the library at Nineveh in the ancient Sumerian kingdom dates to the seventh century BCE. The foundation of the technological society, in fact, was laid in Sumeria more than 4000 years ago at the library of Nineveh.

Monuments are far more than imposing architectural edifices; they are awe-inspiring statements about "who we are." Travel to the moon, for example, is not an economic venture so much as it is a *monument* proclaiming grandeur. The monumental achievement of the Apollo program was not simply to harvest resources or build colonies on the moon (neither of which have happened in the ensuing 50 or so years since Neil Armstrong first set foot on the moon), so much as to proclaim the grandeur of the United States over its chief competitor, the Soviet Union (Psenka, 2008).

One puzzle that we seek to resolve is why civilizations such as China and the Middle East, which did have sophisticated engineered edifices and substantial technical accomplishments, whether the magnetic compass or explosives, did *not* develop the concept of technology. The answer, which we will develop in the final section of

[2] Linguists make a strong distinction between "emic" and "etic" meanings. In *phonetics*, differences of intonation and vocalization are universal, even if not recognized in many languages. The difference between a long "o" ("hello") and a short "o" ("not") is obvious to any ear. Phonemic distinctions by contrast are those that are meaningful within a particular language or discourse community. The difference between "girl" and "gal" is less a distinction of gender and more a distinction of relationship, meaningful more within colloquial English than in other nationalities. Similarly, the etic and emic meanings of technology, in terms of tools as contrasted to the tool assemblages of the Industrial Revolution, did not exist before the word "technology" came into widespread usage in the nineteenth century. Imposing this modernist term on premodern toolkits is, at best, anachronistic, if not ethnocentric. Characterizing a toolkit as "technology" creates additional spaces for exploration and exploitation of the instrumental and social possibilities, whether the application is an internal combustion engine to electrifying homes or the creation of new algorithms for manipulating users' attention on search engines.

this chapter, is that these civilizations were more self-contained and inward-looking, whereas Europe for more than a thousand years was outward looking, from the Norman Conquest to the discovery of the New World.[3] Of course civilizations had since time immemorial been seeking new lands to *colonize*, but not to *incorporate.* The Roman conquests of Gaul or Carthage did not assimilate the south of France or the north of Africa into the Roman Republic, but rather set it aside as a place to which Roman legionnaires could retire. Inevitably the emerging Roman *Empire* became too unwieldy for the Roman *Republic* to manage.

The collision of continents, beginning with the Spanish discovery and subsequent settlement of the New World, created a dynamism that eventually led to the Enclosure Acts[4] and the Industrial Revolution. From the seventeenth century onward, farmlands, (the commons) on which peasants had subsisted, were privatized, evicting peasants off the land and into the growing industrial cities. One consequence of the enclosures was the creation of a large mass of idle laborers who were available to work in the emerging mill towns of Manchester and Leeds, laying the foundation for the Industrial Revolution which is typically dated from the early nineteenth century.

In sum, humanity has been perfecting both its tools and its institutions for millennia, but more typically along single trajectories rather than with complex dynamics. Similarly, human civilizations have been *colonizing,* but not *incorporating* their hinterlands for similar periods of time, but also maintaining a single trajectory. It was only with the Industrial Revolution that humanity began to embrace the dynamism and disruption that are inherent in a technological society. (See Chap. 14 for more on disruption.)

Evolving an Institutional Ecosystem: Guilds, Academies, Libraries, Regulation

Part of the ecosystem of technology is institutional sophistication. As long as tools, whether hammers or horse-drawn plows or metal forges were simple tools, their institutional ecosystem could be and was simple. This institutional ecosystem, in the ownership of farms or factories, and the training of workers, was the basic premise. As tools became more complex, whether gasoline-powered tractors or

---

[3] Although it would require an entire volume to understand this difference in civilizations between East and West, for the moment we can observe that *exploration,* rather than *exploitation*, has been a Western dynamic. Exploration, a fundamental curiosity about the world around us, contrasts with exploitation, the mining of that world. Exploration creates a dynamism that is one of the features of Western civilization.

[4] Enclosure Acts "A series of United Kingdom Acts of Parliament which enclosed open fields and common land in the country, creating legal property rights to land that was previously considered common. Between 1604 and 1914, over 5,200 individual acts were put into place, enclosing 6.8 million acres." https://courses.lumenlearning.com/suny-hccc-worldhistory2/chapter/the-enclosure-act/.

industries building automobiles, institutional complexity was required, whether in the regulation of fossil fuels or highway safety or designing newer conveyances.

Institutional complexity is a central feature of the technological society and its predecessors. Complexity in and of itself embraces levels of authority, demographic diversity, functional specialization, and levels of inequality. In *Complex Organizations: A Critical Account,* Charles Perrow (1986) describes how an *industry* can be made up of multiple entities, public and private. The agricultural *industry*, for example, includes not only family farms, but also agribusiness corporations such as Monsanto, Cargill, and Perdue, trade associations such as the Farm Bureau, and public agencies such as the Department of Agriculture (USDA) and agencies assuring food safety (the Food and Drug Administration). A major anchor of the agricultural industry are the Land Grant universities, first founded in 1862 under the Morrill Act and later expanded to include such institutions as Cornell University and the Massachusetts Institute of Technology (Dodd, 1911). The Land Grant Universities extended higher education to the "agricultural and mechanical classes" in the words of the Morrill Act, an unprecedented democratization of higher education that made the United States a leader in technology. Students come from all over the world to study in American universities, a fact that has benefitted both their home countries and the United States.

Institutional complexity, if not counterbalanced with adequate safeguards, as we elaborate on in Chap. 10, simply creates a more brittle world, prone to collapse. Complex organizations can be more brittle and failure-prone or more resilient and durable depending on the attention and resources invested in them. The entire course of human evolution has been one from increasing complexity. The denial of complexity, whether in the form of the denial of the findings of science or the rejection of racial diversity, is an ever-present feature in the technological society driven by a narrative that projects ease and simplicity while obscuring the complexity that underpin its workings.

The sophistication of tools and institutions coevolved and in fact reinforced each other, whether in the expansion of markets for agricultural produce or the institutional complexity of current financial markets, with the Securities and Exchange Commission (SEC), the Consumer Financial Bureau, the Federal Reserve Bank, and more than a dozen others including the FDIC, the Office of the Comptroller of the Currency, and Financial Industry Regulatory Authority (FINRA), regulating the industry at federal and state levels.

A critical part of the institutional ecosystem is the *public* regulation of technology, an issue that has become a burning debate as technologies race ahead. For but one of numerous examples, social media has become the all-seeing eyes of George Orwell's Big Brother, and the citizens of Orwell's Oceania were always warned that "Big Brother is watching." Nowadays users carry around Big Brother's all-seeing eyes in their pockets on their smartphones, which *records* not only their movements but also their browsing history, shopping preferences, and even "private" conversations. Regulation of the data collected by Big Tech is a major issue in the third decade of the twenty-first century. In the European Union, the General Data Protection Regulation

(GDPR)[5] limits what data companies can and cannot collect, whereas in the United States, Big Tech is mostly unregulated. How, where, and when technological devices can collect and store data on their users is actively debated, a debate colored by the tribalism of contemporary politics, in which (typically more educated) technologists resist encroachment of public agencies on contested private spaces.

A major issue in contemporary jurisprudence is the ongoing the fate of section 230 of the Communications Decency Act of 1996, which shields online platforms from liability for user-posted content. The Communications Decency Act was written at a time before the Internet, when broadcast media used the airwaves. With the new technology of the Internet, the Supreme Court, as Justice Elena Kagan described, "You know, these are not like the nine greatest experts on the internet," Justice Kagan said of the Supreme Court, to laughter (Liptak, 2023). Regulation of Big Tech is today a very hot issue, with an overwhelming democratic consensus that Big Tech has become much too powerful, arrayed against the lobbying might of Google, Meta, and Alphabet. These companies are what Tim Wu (2016) has called "attention merchants"—their business model is one of harvesting users' attention for sale to advertisers. Nearly everyone has had the experience on the Internet of a pop-up ad reflecting their recent browsing history, a feature enabled by monitoring browsing. As consumer products become more specialized and fine-tuned to individual tastes and algorithms zero in on these tastes, the entire *industry* of advertising has fine-tuned its appeals based on users' browsing history, networks ("friends"), and recent purchases. The "attention economy," as introduced by Tim Wu, has a new set of principles far beyond what Adam Smith described for the economy of goods and services.

This is a major step forward in cultural evolution, from simple to complex, from collective to individual. It is in accordance with the progress of civilization, understood not simply as the culture of cities and empires but as an advancement of civility, in stark contrast to the tribalism of earlier times. Tribalism is the designation of an "other" as a people that we have nothing in common with and is an affliction that has beset human societies since the beginning of time. Tribalism today, abetted by technology, specifically social media, can be seen in the antagonism between Christians and Muslims, Protestants and Catholics, and contemporarily ideological camps with the "culture wars" driving contemporary politics. Technology, in fact, can be an ingredient of civility, and advances in civilization and civility have often corresponded with advances in technology. In the reconstruction of our contemporary technological society, we need to find new foundations for civility.

---

[5] The General Data Protection Regulation (GDPR) is the toughest privacy and security law in the world. Though it was drafted and passed by the European Union (EU), it imposes obligations onto organizations anywhere, so long as they target or collect data related to people in the EU. The regulation was put into effect on May 25, 2018. The GDPR will levy harsh fines against those who violate its privacy and security standards, with penalties reaching into the tens of millions of euros. https://gdpr.eu/what-is-gdpr/.

The Plate Tectonics of the Industrial Revolution

A revolution, in contrast to an "evolution," is not simply a gradual change, but an abrupt break with the past, even if tensions had been accumulating unseen for many years. When geologists discovered that the Earth's crust is made up of plates that float uneasily on the earth's mantle and that these plates have been on the move for millions of years, they arrived at an explanation for what had been previously understood as an act of God. When the plates collide, the ground trembles, if not actually collapses. The Industrial Revolution was not simply a perfection of productive tools (factories, looms, and iron forges), but an abrupt displacement of tens of thousands of peasants, a cosmological crisis (see Chap. 8 on cosmological crises) that is not always appreciated as such. Cosmological crises—the world turned upside down—are defining events in human history, redefining "who we are." The convergence of several of these crises that are described below created the conditions that ushered in the Industrial Revolution.

The rearrangement of the institutional ecosystem included not only the displacement of laborers into company towns, but also the decline of guilds, as factories were manned by displaced peasants and village merchants were displaced by itinerant peddlers. It also includes the growing power of nations as the *nation* rather than the kingdom (and the king's tax collector) became more the focus of economic life. It further included a form of economic activity, "capitalism," that created a new dynamic, replacing the stasis of the feudal economy. The fact that all these events happened in a compressed time, roughly from the seventeenth to the nineteenth centuries, suggests a world turned upside down.

As previously noted, the enclosure movement, roughly from the sixteenth to the nineteenth centuries, was a factor leading up to the Industrial Revolution. Millions of peasants were displaced off the commons and into the growing industrial cities of Manchester and Birmingham. Closely related to this was the rise of the factory system in which production took place in a central location with product and process supervision, displacing the authority of the guilds. Contributing events included the rise of colonialism and slavery, as the Age of Exploration (roughly from the fifteenth to the seventeenth centuries) in which Europeans, led by the Portuguese and the Spanish, later followed by the Dutch, English, and French colonizers. The Age of Exploration led to the harvesting of human slaves, primarily from Africa, who fueled the New World economy. The industrial-scale slave trade in the eighteenth and nineteenth century was a tectonic event in the creation of the modern world. Indeed, the scaling up of not just factory and agricultural production but all aspects of production and distribution in the eighteenth and nineteenth centuries turned the world upside down.

An unsung part of the Industrial Revolution is what is sometimes called the "golden age of piracy" roughly from the 1650s to the 1730s, when pirates ruled the seas and in fact established their own principality, Libertalia, on the island of Madagascar. Although pirates are usually thought to be outside the law, as anthropologist David Graeber describes in *Pirate Enlightenment or the Real Libertalia*

(2019), in fact they established a rogue regime (or series of rogue regimes) with their own laws, following the Age of Discovery from the fifteenth through the seventeenth centuries. The pirate regimes, on the borders of civilization (both Christian and Muslim), were in fact a source of dynamism for both.

Also included in seismic shifts that created the conditions for the Industrial Revolution and the technological age were events such as the decline of feudal monarchies, starting with the Bourbons in France and later the Hapsburgs in Austria-Hungary, and the kingdoms of Aragon and Castille in the Iberian Peninsula. And finally, we must include the Protestant Reformation,[6] roughly from the sixteenth to the seventeenth centuries, that upended the authority of the Roman Catholic Church and created numerous and competitive opportunities for worship outside the Church. The connection between Protestantism and capitalism has been debated in numerous sociological texts, most notably Tawney's *Religion and the Rise of Capitalism* (2024) and Weber's *The Protestant Ethic and the Spirit of Capitalism* (2003).

Some of the tectonic plates whose collisions led to the Industrial Revolution included not only continents (both the Americas, as well as the far east explored by Marco Polo, whose explorations were seminal for the making of the modern world), but also consumers and classes. Class distinctions—the difference in productive roles—are uniquely a feature of the *modern* world. Earlier empires obviously had differences in wealth and power, but these were foundational and a matter of status or estate, not negotiable or based on productivity. The enclosure movement, as described, was a tectonic event, where peasants were driven off the land and eventually into the emerging mill towns of Manchester and Leeds. The rise of a class of industrialists was perhaps the greatest disruption that European society had experienced since the Black Death and a precursor for the major disruptions of the industrial society.[7] These industrialists included Richard Arkwright (1732–1792), inventor of the water frame and the factory system and a leader in the Industrial Revolution, or Samuel Crompton (1753–1827), inventor of the spinning mule, a machine for spinning cotton.

Perhaps the initial driver of continental collisions leading up to the Industrial Revolution was the discovery of the Americas, first by Spain at the end of the fifteenth century and later by Britain as itemized in Table 2.1 in the sixteenth and seventeenth centuries.[8] This was not simply a discovery of new territories, but an opening of the European mind: the world was far larger and more complex than the Tudors and

[6] "The Protestant Reformation that began with Martin Luther in 1517 played a key role in the development of the North American colonies and the eventual United States." https://education.nationalgeographic.org/resource/protestant-reformation/

[7] There is substantial literature exploring the upside and downside of disruption. While many have celebrated the disruption of the technological society (Christensen, 1997), others, particularly at the bottom rungs of the social ladder have found their lives more miserable.

[8] There is archeological evidence that Norse-Icelandic explorers Lief Erickson and Bjarni Herjolfsson discovered Vinland believed to be on the North American continent, predating the arrival of Columbus by 500 years (Kuitems et al., 2021). Although the claim is that there were subsequent visits to Vinland by Norse explorers, it was not established as a colony.

Table 2.1 New world colonial expeditions

Explorer	Discovery, actual or intended	Dates
Christopher Columbus	Indies	1492
Ferdinand Magellan	Circumnavigation of the globe	1519
Vasco de Gama	Rounding the Cape of Good Hope	1524
Sir Francis Drake	Circumnavigation of the globe	1577
Sir Walter Raleigh	British colonization of North America	1584
Hernando Cortes	Fall of the Aztec Empire	1519–1520
John Cabot	Coast of North America	1497
Samuel de Champlain	Settlements in Canada	Seventeenth century
Martin Frobisher	Northwest Passage	1577–1578

the Bourbons and Medicis had imagined. A diversity of languages, religions, and cultures confronted the Europeans.

What to do with these new land masses was a dilemma that confronted England, Spain, Portugal, Italy, Belgium, the Netherlands, and France. Each of these nations took a unique trajectory, whether the colonization of the New World by England, Spain, and Portugal, or the enslavement of African people by Belgium and England. The "Age of Discovery," running roughly from the fifteenth to the seventeenth centuries took European powers all over the world, opening the European mind. The resolution of this dilemma was that these countries took different trajectories, but all became the engines of a new emerging economy. Part of this new emerging economy became the Industrial Revolution, dating approximately from the mid-eighteenth century to the early nineteenth. Large sections first of England and later of continental Europe built factories, employing displaced peasants, and supporting an emerging consumer economy.

Out of the Industrial Revolution came a new concept, technology, dignifying the craft or *techne* of industrial workers with *logos*, the authority of the written word. This word, "technology," which did not exist before 1621 (its first mention in the Oxford English Dictionary) yet not in common usage before Bigelow's, 1829 *Elements of Technology* at the outset of the Industrial Revolution. "Technology" turned the world upside down, no less than continental collisions of tectonic plates turned village lives upside down.

Economic Diversity

One consequence of these tectonic collisions is a greater economic diversity than had prevailed in the Middle Ages. As is well known, a diverse economy is more creative than a monoculture, and the opening of the European mind created an opening for new and diverse forms of production and distribution which eventually resulted in the Industrial Revolution.

An important dynamic of the Industrial Revolution was a diversification of the economy, first in England and later throughout Europe. The Industrial Revolution, which made Europe, and later the Americas, Asia, and ultimately Africa a more prosperous and dynamic, is intimately bound up with a central dynamic of the ongoing prosperity, technology. Technology creates an economic diversity that neither agriculture nor industrialization can support.

Beginning in the sixteenth century a new emphasis on private goods began to grip Europe. The stable arrangements of the feudal society of the Middle Ages were upended both by the corruption of the feudal order under emperors such as the Tudors and the Bourbons and the Hapsburgs and by the rise of an emerging class who saw new possibilities in the privatization and enclosures of what had previously been public lands. Adam Smith's, 1776 celebration of private goods in *The Wealth of Nations* was the culmination of a centuries-long erosion of the common good.

In sum, the emergence of "technology," as we detail in the next chapter, is neither a natural evolution nor an historical inevitability. Rather it was something, much like technological wonders such as air travel and nuclear power, that had *intentionality* and identifiable human and nonhuman actors behind it. Whether these intentions were benign or malignant, and whether these actors were heroes or villains are questions that have been debated ever since. Manchester millhands, for example, saw in the "dark Satanic Mills" that Blake described as a form of enslavement. William Blake's *Jerusalem*[9] gave what was probably the first literary mention to the emerging mills of the Midlands:

> And did the Countenance Divine,
> Shine forth upon our clouded hills?
> And was Jerusalem builded here,
> Among these dark Satanic Mills?

By contrast, the industrialists such as Arkwright and Crompton saw themselves as heroes bringing greater prosperity to the world. Fast forward two centuries and we are debating whether Google is liberating or enslaving humanity, and whether Amazon's "fulfillment centers" are more simply a twenty-first century dark Satanic Mill, only with improved lighting and air conditioning. The emperors of the technological society, as we describe in Chap. 11, have amassed powers, hidden behind flat-panel screens, far greater than that of Nero or Caligula or the Bourbons, yet are worshiped or at least admired by millions.

"Power corrupts," Lord Acton famously said, and "absolute power corrupts absolutely." The corruption of power can be seen throughout history, whether in the failures of the Bourbon monarchs in eighteenth century France or the excesses of capitalism in America's Gilded Age in the nineteenth. The pursuit of corruption by today's technology giants signals not simply venality of an individual or an industry, of a Musk or Bankman-Fried or a Zuckerberg, but the decline and fall of an entire

---

[9] https://www.poetryfoundation.org/poems/54684/jerusalem-and-did-those-feet-in-ancient-time

civilization. *Corruption*—the slighting of worthier goals for individual base pleasures and gratifications—is not simply a moralistic scolding, but an actual thermodynamic debasement, substituting smaller objectives for larger. The legacy of Mark Zuckerberg and Elon Musk may be in celebrating corruption but also in glorifying the decline and fall of the entire modern construction of a technological society.

In conclusion, we can see that the building blocks of technology, like the bricks and beams in the construction of a cathedral, were all present at the outset of the Industrial Revolution. Assembling the edifice that we now label as technology, a word that was only coined in the seventeenth century, is the subject of our next chapter.

References

Bigelow, J. (1829). *Elements of technology: The useful arts considered in connexion with the applications of science*. Harper.

Dodd, W. E. (1911). [Review of The Origin of the Land Grant Act of 1862 and Some Account of Its Author, Jonathan B. Turner, by E. J. James]. *American Journal of Sociology, 17*(3), 406–407. http://www.jstor.org/stable/2763175

Douglas, M. (1986). *How institutions think*. Syracuse University Press.

Frontinus, S. J. (2004). *De aquaeductu urbis romanae*. Cambridge University Press.

Graeber, D. (2019). *Pirate enlightenment, or the real libertalia*. Farrar, Straus, and Giroux.

Liptak, A. (2023). Supreme court seems wary of limiting protections for social media platforms. *New York Times*.

Pollio, M. V. (2020). *The ten books of architecture*. Independently published.

Psenka, C. E. (2008). *A Monumental Task: Translating complex knowledge in NASA's human space flight network*. Wayne State University.

Smith, A. (1776). *The Wealth of Nations*.

Tawney, R., H. (2024). *Religion and the rise of capitalism: A historical study*. Independently Published.

Weber, M. (2003). *The protestant ethic and the spirit of capitalism*. In T. Parsons (Ed.). Dover Publications.

Wu, T. (2016). *The attention merchants: the epic scramble to get inside our heads*. Vintage.

Chapter 3
The Constitution of "Technology"

Abstract In this chapter, we seek to unpack the multiple social, physical, and conceptual elements that in the eighteenth and nineteen centuries were assembled to constitute technology as we understand it today. Far from being a natural evolution, "technology" burst forth on the world stage as a deliberate *construction* of a physical, social, and *institutional* edifice. The institutional foundations of technology we suggest are no less important than its material or cybernetic foundations. Unpacking the meaning of "technology" in terms of its multiple physical, social, and institutional facets is the objective of this chapter.

What is "Technology"?

From Jacques Ellul's *The Technological Society* (1947) onward, there has been substantial literature questioning the nature and benefits of technology. Ellul's title in the original French, *La Technique ou l'enjeu du siècle,* the inevitability alluded to by *l'enjeu,* "the challenge," suggests a fatalism that perhaps is at the heart of the technological society. Interest in technology followed major events of the 1940s, both the horrors of the Holocaust, which industrialized mass extermination using sophisticated devices to kill and incinerate millions, and the explosion of nuclear bombs over Hiroshima and Nagasaki, using the latest scientific advances. The rapid technological innovation of the 1950s and 1960s brought new devices from refrigerators to televisions to automobiles to microchips into the lives and households of millions, thereby convincing millions of commentators that the Golden Age of Technology had arrived. While some such as Ellul condemned, or at least questioned, the benefits of technology, for most there was an unquestioned assumption that technology was a Good Thing.[1]

[1] At the risk of pedantry, we must make it clear from the outset that within the engineering and social science literature and popular usage there are two contrasting and possibly inconsistent usages of the word "technology." The more commonplace usage is to designate as "technology" any sophisticated tools such as computer software or internal combustion engines. From this perspective, eating utensils and hand axes are not "technology" unless it is qualified with a marked term "primitive technology." Within the social science literature on the other hand any culturally specific toolkit

A. Batteau and C. Z. Miller, *Tools, Totems, and Totalities*,
https://doi.org/10.1007/978-981-97-8708-1_3

What is this Good Thing? Is it a tool of the Devil, designed to enslave us all, much like Emerson's observation that "things are in the saddle and ride mankind" (1847), or is it a magic wand that can move mountains, create new jobs, and relieve human drudgery? We have argued that the concept of "technology," the joining of *techné* (craft or skill) with *logos* (the authority of the written word), is a very recent invention, appearing (according to the Oxford English Dictionary) for the first time in 1621 as we noted in the previous chapter, at the outset of the enclosure movement in England. Unearthing the historical heritage of this Good Thing is a first step toward imagining a society beyond the modern construction of hegemonic technology, a society in which relationships count for more than instrumentalities.

In this chapter, we seek to unpack the multiple social, physical, and conceptual elements that in the eighteenth and nineteenth centuries were assembled to constitute technology as we understand it today. Far from being a natural evolution, "technology" burst forth on the world stage as a deliberate *construction* of a physical, social, and *institutional* edifice as described below. The institutional foundations of technology we suggest are no less important than its material or cybernetic foundations. Unpacking the meaning of "technology" in terms of its multiple physical, social, and institutional facets is the objective of this chapter.

Historical Intersections

There were multiple historical developments from the sixteen to the nineteen centuries that in their native state might seem disconnected. In fact, like the construction materials of wood, stone, glass, and fabric were *associated.*[2] To create a cathedral, these historical developments came together in the seventeenth and eighteenth centuries to create the edifice of technology as we understand it today. Leading the field was the discovery of the New World. Prior to 1492, Europe was a world without a West. Although Pythagoras had demonstrated that the world is round, not flat, when Christopher Columbus dared to sail through the Pillars of Hercules, he expected to reach the Indies, instead he found the New World. This continental collision of Europe and Asia with what came to be called the Americas was a bringing together of new cultures and economies that have been a source of dynamism and disruption ever since.

This disruption was a cosmological event, a reordering of the shape of the universe, in which people began to question not only "Where are we?" but also "Who are we?" (See Chap. 8 for further discussion of cosmological crises.). Europe, which had been at the center of the world (and the universe), with outliers on its subordinate

is a "technology." This compression of multiple contrasting and sometimes inconsistent usages is typical of key words (see below, ""Technology" as a Key Word").

[2] The use of "associated," meaning to bring together in an intentional way, is a reference to John Law's concept of association and dissociation in "Technology and Heterogeneous Engineering: The Case of the Portuguese Expansion" (Law, et al., 2012).

fringes, was now dethroned, having to *share* the globe with peoples who never heard of the Judeo-Christian God. The resulting humility was an initial spur to the Age of Exploration. *Exploration* implies an acceptance of complexity and uncertainty, a curiosity and a humility, an awareness that there is more to discover beyond the homeland and people. The Age of Exploration, roughly the sixteenth and seventeenth centuries, expanded Europeans' (and later Asians') comprehension of the world and its possibilities. An *acceptance* of exploration and dynamism has been a characteristic of technological societies ever since.

The anthropologist Eric Wolf, in *Europe and the People Without History* (1982), examined how in the premodern era European colonialism created a "tributary economy" with raw materials (including slaves) flowing into the metropoles of Europe. This expansion of trade on a global scale eventually resulted in the Industrial Revolution. Far from being driven by inventions such as the cotton loom or the steam engine, the Industrial Revolution was a consequence of European expansion into all corners of the globe. New flows of slaves, cotton, and millhands eventually knit the world together with England as the metropole.

Equally important were three inventions that transformed the world: gunpowder, the compass, and paper and printing. Gunpowder, which was invented by the Chinese in the first millennium, ultimately transformed the nature of warfare and relations. Prior to this, armies fought with spears and swords, but could not kill at a long distance. Gunpowder, which the Chinese used in fireworks, was an accidental discovery and used ceremonially for many years before it was weaponized. This contrast between ceremonial or ornamental use on the one hand and practical use on the other is an important theme for us: the Internet, developed in the 1980s, was for the first ten years of its existence *not* a vehicle for commerce or warfare, until the 1990s; the first e-commerce sites in 1995 and the first bots in the late 1990s commercialized and weaponized the Internet. With the coming of gunpowder (and cannons and muskets), warfare changed significantly and could be scaled up to a continental level. Aside from use in ceremonial fireworks, the Chinese kept gunpowder under strict control.

Equally notable was another Chinese invention, paper and printing. Prior to the introduction of paper and printing to the West in (roughly) the fifteenth century, the written word was reproduced by scribes on parchment. The limitations on the spread of reading are obvious. Beginning with what many have called the Gutenberg Revolution (printing with moveable type) in the fifteenth century, mass literacy became a possibility. This eventually led to the Age of Nationalism and the "imagined communities" of nations that reordered the world.

Prior to the fifteenth century *communities* were structured around cities and churches, sites where face-to-face relationships both between equals and between superiors and subordinates were the only option. With the spread of mass literacy, citizens residing in far corners of the kingdom could *imagine* themselves belonging to the same community, creating the imagination of nations and the age of nationalism, far-flung communities clustered around a common mythology. A digression on the history of nationalism would carry us away from our central theme of the constitution of technology, except to note that the modern construction of technology and the rise of nationalism were roughly coincident.[3]

The links between the modern construction of technology and the construction of nations and national identities need to be unpacked. These two events, roughly coincident, are in fact mutually reinforcing on multiple levels. Perhaps the most obvious are the technologies of nation-building, whether transportation technologies (roads and railways) or communication technologies (the press and later mass media). Less obvious, but no less important is the fact that standards are national creations. The differences between European and American standards for electrical current, weights and measures, or the width of highways, are examples of this. Standards are created by national bodies, and the approximately 160 standards organizations around the world, including American National Standards Institute (ANSI) and British Standards Institution (BSI), attest to this. At a more profound level the very idea of standards is an expression of national identity. National identity, the sense that we are "one people" with a common ancestry in the mists of a distant past, is very much a modern creation, and nations are the epitome of "imagined communities." Medieval guilds, prior to the rise of the nation-state, had their own standards, included in the "mysteries" of the craft guilds, which apprentices had to learn. With the rise of the nation-state came the rise of national standards, not only for electric power but also for such mundane matters as the pitch of screw threads and the configuration of electric outlets, so that technological standards and national identities can be seen as mutually reinforcing. There is also a *spiritual* aspect to standards, the fact that they inspire the nations that support them is no less important than their technical aspect. This inspiration is less a spiritual depth and more a unifying moment in the invention of *anything*.

A further Chinese invention that transformed the world was the magnetic compass. The Chinese discovered magnetism during the Han dynasty, 206 BCE to 220 A.D., and soon realized that the Earth's magnetic field could be used for navigation. In the West's "Age of Discovery," with navigators beginning with Columbus and going through Magellan and Amerigo Vespucci explored and expanded the known world.

As noted in Table 3.1, some of the major European adventures in Africa and the New World include not only the voyages of Columbus, but also Magellan, Vasco da Gama, and Martin Frobisher. Frobisher, looking for the "Northwest Passage" to the Indies, was the first known European to set foot in what is now Canada, and Vasco da Gama, rounding the southern tip of Africa, demonstrated that one need not travel overland to reach the Indies. We characterize these events as "tectonic," a collision of continents, less in terms of the physical landmasses and more in terms of a collision of unfamiliar societies and cultures.

The cosmological reorientation left Europe open to reimagining the world. The opening of the European mind has been a source of dynamism around the world ever since. A theme that we shall return to is the idea of a cosmological crisis—"the world turned upside down." Examples of cosmological crises in history include the

[3] See Benedict Anderson's *Imagined Communities* (1983) on the construction of nationalism.

Table 3.1 European explorations

Explorer	Discovery, actual or intended	Dates
Christopher Columbus	Indies	1492
Ferdinand Magellan	Circumnavigation of the globe	1519
Vasco de Gama	Rounding the Cape of Good Hope	1524
Sir Francis Drake	Circumnavigation of the globe	1577
Sir Walter Raleigh	British colonization of North America	1584
Hernando Cortes	Fall of the Aztec Empire	1519–1520
John Cabot	Coast of North America	1497
Samuel de Champlain	Settlements in Canada	Seventeenth century
Martin Frobisher	Northwest Passage	1577–1578

collapse of the Roman Empire and the waning of the Middle Ages. With the dawn of the industrial society the world was continually turned upside down.[4]

A historical puzzle that has occasioned substantial commentary is why these three or four inventions, which arguably transformed the West, yet originated in China, did not have a similar transformative impact in China. This puzzle, which is sometimes referred to as the "Needham paradox," has been the subject of considerable commentary (Jacobsen, 2013; Lin, 1995; Needham, 1954), which can be (perhaps inadequately) summarized here as a preference in Chinese culture for stability, and a European openness to exploration and experimentation. China was the center of the universe, so why explore elsewhere? What was there of value beyond the Middle Kingdom? Although ancient kingdoms explored their hinterlands for millennia, whether the Romans in Gaul or the Chinese in their wild west, these explorations were always on the fringes and never a part of the central dynamic of the kingdom. Roman legionnaires imagined retiring to the south of Gaul, but never building a civilization there; similarly, Portuguese and Spanish and Belgian explorers saw the continent to their south as a place to harvest raw materials including humans for enslavement, rather than a place to extend their kingdoms. Ancient empires including the Roman and the Macedonian did not *imagine* dominating the entire globe, a concept which did not even exist until the modern era. Only with the Renaissance and the opening of the European mind did exploration and empire-building become a central dynamic to French, English, Portuguese, and Spanish cultures.

One final transformation that gave rise to the Industrial Revolution was the enclosure movement in England, beginning (roughly) in the seventeenth century. As noted in the previous chapter, open or common fields, on which peasants had raised their livelihoods for centuries, were taken away and appropriated by the aristocracy, thus creating a substantial impoverished landless population available to labor in the emerging industrial cities of Manchester and Lancaster. This was the foundation

[4] When the colonial army defeated the British at the Battle of Yorktown in 1783, marking the end of the Revolutionary War, the British band played "The World Turned Upside Down," an appropriate comment for the defeat of an imperial power by its colonial subjects.

of the modern economy and industrial society, which many have celebrated. The Industrial Revolution made it possible to scale up inventions such as the printing press to become world-transforming devices, notably altering the cosmology of the Europeans and later the entire world.

The *corruption* of the feudal system represented by the enclosure movement is notable. One need not make ethnocentric judgments to conclude that within some groups and situations there is corruption. Every society makes judgments about goods that are noble and ignoble, and corruption is the deliberate slighting of higher goods for the less noble. Examples of such corruption would include the neglect of family for individual pleasure or the undermining of the larger community for private advantage. Notable examples of corruption throughout history would include cover-ups and denial of sexual abuse in large organizations such as the Catholic Church or the private appropriation of public funds covered up by political parties. The "cover-up" in public life is notable: corruption is not simply an individual vice but rather a collective flouting of larger goods. Corruption is a significant precursor to social revolutions, whether the corruption of the feudal order at the close of the Middle Ages or the corruption of the Bourbon monarchy prior to the French Revolution. The corruption of today's hegemonic technological society, in which the technological society is more focused on tools than on their ultimate purposes, should similarly suggest an impending social revolution.

Finally, we should take note of the major trend in western literature of this period, the emergence of Romanticism. It might seem odd to link the constitution of technology to Romanticism, but for the fact that technology has offered a vision of liberation familiar to Romantic poets such as Wordsworth and Shelly. Blake's "Auguries of Innocence"[5] sums up the magical transformations afforded by technological implements:

> To see a world in a grain of sand,
> And heaven in a wild flower,
> Hold infinity in the palm of your hand,
> And eternity in an hour.

Romanticism, as a reaction both to enlightenment rationalism and industrial regimentation, afforded a vision of perfect freedom that technology nowadays also affords. Technology, considered in the abstract, offers the world in a grain of sand and eternity right now. In the words of David Lilienthal, the founder of the Tennessee Valley Authority,

> There is almost nothing, however fantastic, that (given competent organization) a team of engineers, scientists, and administrators cannot do today. … When [builder and technicians] have imagination and faith, they can move mountains; out of their skills they can create new jobs, relieve human drudgery, give new life and fruitfulness to worn-out lands, put yokes upon streams, and transmute the minerals of the earth and plants of the field into machines of wizardry to spin out the stuff of a way of life new to this world. (Lilienthal, 1945, pp. 94-96)

[5] https://www.poetryfoundation.org/poems/43650/auguries-of-innocence Source: *Poets of the English Language* (Viking Press, 1950).

In sum, visions of perfect freedom and domination over nature are afforded by technology and the technological imagination, places infinity in the palm of one's hand. The modern *construction* of technology is not simply the construction of artifacts which many have commented on but also the construction of the entire concept and expectations of technology, its wonders, and its components including artifacts and conceptual standards and national aspirations. This convergence of artifacts and national aspirations is very much a modern construction coincident and conceptually convergent with multiple aspects of the modern world. The tension between the domination of nature and living in harmony with nature has been with us since the Industrial Revolution.

Construction v. Emergence

The usual narrative of technology is that it *emerged* out of the scientific revolution and the enlightenment. "Emergence" suggests an undirected evolution with no guiding hand. By contrast, "construction," which we argue for here, suggested a *directed* and *intentional* process, with a coherent vision of the ultimate purpose. A major thread, the "social construction of technology" (Pinch & Bijker 1984; Bijker et al., 1987; Winner, 1993), states that technologies are *constructed* in response to problems experienced by relevant social groups, through a process of interpretive flexibility and leading to closure and stabilization. The steam engine, for example, was invented in response to the pressures of production in the emerging factories of the English Midlands, just as Whitney's cotton gin, invented in 1793, was in response to the slave economy of the American South. As enslaved people harvested increasing quantities of cotton, a production bottleneck, removing the seeds from cotton bolls, was solved by Eli Whitney in 1793.[6] Technologies eventually result from a process of *closure*, the end of debate specific to a topic or event, and *stabilization*, including the definition of standards, that fix the technology. Marx wrote that the steam engine gave us the factory system, whereas the historical sequence suggests that the factory system (which predated James Watt's invention by several centuries) gave us the steam engine. Technological innovations, ranging from the steam engine to nuclear power to computers, are constructed to resolve problems that changing social relationships created.

In other words, the modern *construction* of the concept of "technology" as contrasted to earlier tools is less an evolution and more a guided intentional process, in which *logos* is just as important as *techné*. The guiding hand for the modern construction of technology was the up-and-coming class of industrialists in the English Midlands. The emergence of industry and the modern construction of the concept of technology went hand-in-hand.

A standard anthropological perspective on the relationship between tools and the course of humanity is presented by Richard L. Currier, in *Unbound* (Currier,

---

[6] https://www.archives.gov/education/lessons/cotton-gin-patent.

2015), which examines eight key devices that profoundly altered the relationship between the human species and the earth's environment. These implements included fabricated spears and digging sticks, which had a demonstrable effect on human anatomy (placing more emphasis on the hands and upright posture), the technology of communication and agriculture, transportation, and in the past five hundred years the technology of precision machinery and most recently the technology of digital information. Currier's broad sweep, consistent with an anthropological perspective that locates *homo sapiens* among the multiple species that inhabit the earth.

We locate technology, in contrast to earlier toolkits, as uniquely associated with the Industrial Revolution. As we elaborate in this chapter, the word "technology" (the joining of skill, or *techné*, with *logos*, the authority of the written word) was only coined in 1621 and only came into broad circulation two centuries later. As Louis Dumont argued with respect to the word "caste," applying the term indiscriminately to cultures outside its original provenance is misleading.[7] Further, "technology" is associated with a set of attitudes, including awesome powers, progress, and man's mastery over nature (the gendered language here is intentional) all of which contribute to the *poetry* of technology, attitudes, and expectations that are not associated with simpler toolkits.

Currier's Promethean perspective of technology out of control, in sum, is orthogonal to the question we are raising about the place of "technology" (and not, more generally, tools) in the contemporary world. Technology, we suggest, is a specific *institution* and not simply an aggregation of instrumentalities, specific to the modern world. As we outgrow our enchantment with technology, we can begin imagining a world beyond hegemonic technology.

Five elements make up the idea of social construction, not only of technology but also any social construct, whether gender, race, or social class. The first of these is *interpretive flexibility*, the idea that any artifact or practice can have multiple meanings. For example, are cars about personal mobility, economic efficiency, or desire? Or is the genius of the car and its display derived from the fact that it embraces all of these, plus multiple additional meanings. In the twentieth and twenty-first centuries, the car is about display, personal freedom, mobility, courtship, heritage, and the domination of industry. Progenitors of this, whether Henry Ford or Gottfried Daimler, are celebrated in museums around the world. Part of interpretive flexibility is *design flexibility*, the fact that any artifact has multiple design options, and these are expressive of different social objectives.

Closely related to this is the fact that *relevant social groups* determine the nature of any technology. For airplanes, are the relevant social groups the military, or business travelers, or tourists, or some combination of all of these? For automobiles are the relevant social groups families, city planners, or shippers? Obviously, some

---

[7] As Louis Dumont (1970) argues in "Caste, Racism, and Stratification," the Hindu caste system, where people are assigned at birth to the caste of their parents, and subsequently marry into the same caste, extrapolating this term to an individualistic society is an ethnocentric imposition of ideas of ("natural") hierarchy onto distinctions that are fundamentally racial.

compromise – a political adjustment – is necessary among all of these, and political dynamics are no less important than the laws of physics to the character of *any* technology. Following these political dynamics is the fact that powerful groups often have a guiding hand in determining the character of *any* technology.

Part of the dynamic is the wider *social context* of the technology, whether a city, nation-state, or an entire continent. What groups and interests have a hand in the development and the meaning of a technology? Who "owns" the car—city planners, corporations, families, or an entire nation? "Ownership" obviously has multiple layers of meaning, not only legal rights as "private property" but also as a club good (expressing identity), and a public good, knitting together a metropolis or an entire nation. *Ownership* obviously varies across multiple cultures, in some even absent. In America, more than any other nation, the car has shaped the community, and the community has shaped the car.

Finally, the *technological frame*, the narrative framing of the meaning of the technology, is typically derived or at least consistent with the other stories that a society tells about itself. Is our society fallen from a prehistoric Golden Age, or is it part of an onward and upward march of progress? The "idea of progress," as Bury in his classic *The Idea of Progress* (1921) notes, is a very recent invention, associated with the Reformation and the Renaissance; previously the narratives of society told of descent from a golden age. Technology offers the narrative of continuous improvement. As we note in the previous chapter, William Blake's "Jerusalem" (1804) is perhaps the best example of this:

Bring me my Bow of burning gold:
Bring me my arrows of desire:
Bring me my Spear: O clouds unfold!
Bring me my Chariot of fire!

I will not cease from Mental Fight,
Nor shall my sword sleep in my hand:
Till we have built Jerusalem,
In England's green & pleasant Land …

One further example would be Matthew Arnold's "Progress" (1822):

The Master stood upon the mount, and taught.
He saw a fire in his disciples' eyes;
'The old law', they said, 'is wholly come to naught!
Behold the new world rise!'

'Was it', the Lord then said, 'with scorn ye saw
The old law observed by Scribes and Pharisees?
I say unto you, see ye keep that law
More faithfully than these …'

As noted above, Romanticism is an important part of the framing of technology. The ultimate purpose in this view would be the improvement and scaling up of industrial production and the enrichment of a rising class of industrialists. Some might argue "well, of course," but the English class system as late as the eighteenth and nineteenth centuries did *not* place a premium on upward mobility; "knowing your place" was a virtue for all classes, and "class jumpers" were frowned upon. The rising class of industrialists, class jumpers all, did not just change the English economy. They also changed the English cosmology, their sense of the order of the universe.

Our point is that not simply the construction of technological implements (steam engine, cotton gin, etc.), but the entire ensemble of technological expectations can be analyzed in terms of relevant social groups, interpretive flexibility, and closure and stabilization. These expectations include moving mountains, creating new jobs, relieving human drudgery, and creating human progress. The relevant social group is the class of industrialists emerging out of the Industrial Revolution, and the interpretive flexibility pertains to how different implements can be used, whether for personal transport or status display or industrial haulage. Closure and stabilization refer to the formation of standards. The emergence of the modern conceptualization of "technology" as standards-based tools that would improve the world gave interpretive closure to the trajectory of the Industrial Revolution.

Key Perspectives: Networks, Standards, Modernism, Diversity

There are multiple perspectives that inform the constitution of technology. At the top of the list is the emergence of large-scale networks, whether networks of trade or military conquest. With the discovery of the Americas, the European world was scaled up, embracing a polyglot of cultures, languages, and religions, this diversity has been a source of dynamism ever since. For but one small example of the importance of diversity, America's technological leadership has been on the shoulders of immigrants and children of immigrants, whether Henry Ford (son of Belgian and Irish immigrants) or Sergei Brin (cofounder of Google and son of Russian immigrants) or Alexander Graham Bell (son of Scottish immigrants). Polyglot and complex adaptive systems have a dynamism that monoglot cultures lack.

We can extend this insight to other systems and ensembles, whether literary or musical or industrial. In music, the contrast between the well-ordered sonatas of Bach and the competing voices of doo-wop (a rhythm and blues style combining jazz roots with rock and roll) is a contrast between order and innovation. In industry, the contrast between well-ordered factory production and improvised industries such as the software industry that more typically started in a garage for example Hewlett-Packard and Apple than in a corporate laboratory.

Vital to the emergence of large-scale networks was the emergence of standards, which we note is a defining feature of technology. Standards, whether the standards for the width of highways or the thread of nuts and bolts, are an essential part of any technology to become more than a local tool. Standards were initially created by craft guilds and then evolved into national standards as industry scaled up. It is worth noting that standards originate as a response to state-level formations which then become international standards as industry scaled up. Examples include standards for electrical current, the width of roads, or domestic plumbing.

A further perspective on technology has been its embrace of modernism, an artistic and literary movement that emerged in the late nineteenth and early twentieth centuries as a reaction to the (static) portrayals of classical genres. The contrast between a Picasso portrait and a classical portrait attests to this (Figs. 3.1 and 3.2).

"Modernism" as an artistic and literary movement is closely related to "modernity," a sociological feature of contemporary society. "Modern society" is characterized by the fleeting experience of urban life. Modernity includes secularization and such (relatively) contemporary philosophical trends as Marxism and post-structuralism. As described in Marshall Berman's *All that Is Solid Melts Into Air* (1982), the twin threads of modernism and modernity are wrapped around a Faustian vision of humanity as the master of the universe.

Fig. 3.1 Classical portrait

Fig. 3.2 Picasso portrait

The result of all of this is what Bryan Pfaffenberger (1992) has called "techno-fetishism," an idealization of (modern) technology as a solution to the world's ills. Fetishism is the attribution of human characteristics to nonhuman objects. It is the treatment of nonhuman objects (both material objects and inanimate beings and concepts, including images and brands) as sentient beings, capable of forming relationships and attachments. Karl Marx, in his *Critique of Political Economy* (1904), described commodity fetishism as the attachment to (mass-produced, market-exchanged) commodities as foundational to a capitalist society.

Fetishism, of course, is an ancient feature of human cultures, and many (perhaps all) tribal cultures have fetishized natural species and inanimate objects for centuries. What is new in the technological society is the fetishizing of (supposedly) utilitarian objects, manufactured not simply for their usefulness, but also the way they can embody connections (or disconnections) among human tribes (Figs. 3.3 and 3.4).

Examples of technological fetishes include guns and automobiles and other techno-totems with which modern societies define identities. Fetishism is the treatment of human-made objects as having supernatural powers. Examples of fetishism in history include religious objects such as relics and idols and priestly vestments. Karl Marx described "The Fetishism of Commodities" as a trait of capitalist societies in which industrially produced commodities were given magical or supernatural powers. Contemporary examples of commodity fetishism would include any goods or services that are exchanged in the market, a transactional connection, contrasted to those that are *shared*. The dissolving of social connections in the icy bath of egotistical calculation (in Marx's words) is a central dynamic of the technological society. One of numerous examples, Facebook's monetization of "friendship," is used for

Fig. 3.3 Paleolithic fetish

Fig. 3.4 Contemporary fetish[8]

---

[8] 'It's who they are': gun-fetish photo a symbol of Republican abasement under Trump | Republicans | The Guardian.

targeted advertising. "Friends" on Facebook are created by algorithms and may well be bots, rather than living, breathing homo sapiens.

Representations of Instrumentality

"Technology," as we have argued, is not a human universal. Applying this term indiscriminately to any toolkit, improvisation, or instrumentality is an ethnocentric imposition, in much the same spirit as imposing Western ideas of class or stratification on non-Western societies (Dumont, 1970). What distinguishes modern technology from earlier toolkits and useful devices is the role of *autonomous representations* in its creation, circulation, use, and evolution. Autonomous representation—images, documents, standards, and instructions not embedded in the devices—creates new forms and new circuits of movement for technology, creating dynamics for evolution and change unavailable to earlier instrumentalities. Although Western architectural documents are centuries old—we might consider Vitruvius's *De Architectura* or Frontinus's *De Aqueductae Romanae* as proto-documentation—those from antiquity are more illustrative and promotional rather than a central part of a design process. Only with the architectural achievements of cathedrals in the Middle Ages (Turnbull, 1993), did autonomous representation become critical to the technology.

Autonomous representation can consist of formalized plans, drawings, blueprints, or written instructions for the construction, use, and maintenance of a technological device, whether a cart or a cathedral. Autonomous representations also include promotional documents and displays. Autonomous representations are unlike the instructions of a master craftsman to an apprentice, which are embodied in the master's voice and quite frequently his hands. Written and pictorial instructions can be severed from the act of stonecutting or woodcarving, enabling many others to imagine the act as well, even if embodied instruction from the master is ultimately necessary for the act. With the rise of print capitalism (Anderson, 1983) in the early Modern Era, opportunities for autonomous representation grew rapidly, and with the graphic revolution (Boorstin, 1972) autonomous representations soared above the rude facts of mechanicals' work. Autonomous representation joined *logos* with *têchne*¸ in ways that they had not been joined in earlier eras.

Autonomous representations can be associated with the rise of cities in the Neolithic period, particularly with the rise of writing. In classical antiquity, however, whether in Qin China or the Roman Empire, reading and writing were skills enjoyed only by a small élite. Technical accomplishments, whether cooking or stonecutting, were typically beneath the dignity of those who could read and write. A document such as *De Architectura*, with its description of walls, roads, dwellings, and temples, written for his patron the emperor Augustus, was intended more for those who commissioned roads and walls than for those who actually built them. Vitruvius can be seen as laying down standards—*De Architectura* combines prescriptive and descriptive language—and in a sense is placeless and timeless.

The considerable body of scholarship on science and technology in ancient China reinforces these conclusions that for systems to evolve beyond local adaptations and instrumentalities, they require some form of formal representation. As early as the Han dynasty one can find Chinese drawings of water mills and horse collars, as well as clocks and crossbows. These representations eventually became the basis for the well-known spread of Chinese technology into and revolutionizing Western Europe, as exhaustively documented by Joseph Needham (1954).

Such placeless and timeless documents are required as instrumentalities evolve beyond local adaptations into large-scale systems and assemblages. Early tools, whether for cutting, pounding, or many other operations, were locally produced, used, and adapted to local conditions and materials. As systems and assemblages scaled up, there needed to be some basic agreements on their materials, design, and functionality. The width of a road, or the height of a wall, cannot be left to the imagination of the local builder or the plurality of conventions in multiple villages. From this necessity, associated with the increasing role of the church and the state in civil projects, came the first standards of measurement and construction, whether for dimensions or materials.

In this respect the construction of cathedrals such as Chartres can be seen as an axial stage, in which an assemblage of multiple local practices were integrated and assembled into a monumental structure (Turnbull, 1993). The templates used by the cathedral builders to reconcile various stonemasons' customary practices were autonomous representations, separable from both the stones and the act of stonecutting. Later, as the achievement of Chartres was replicated, standards evolved out of the local practices of the cathedral builders. Likewise, the templates used by stonemasons building Aztec pyramids could be considered a form of autonomous representation.

Such standards create zones of autonomy within which technologies, whether wall-building (Vitruvius in *De Architectura*) or power distribution (Hughes, 1983), can grow and flourish. They compress the interests and values of multiple parties into coherent, discursive statements. The long-standing debate over technology as an autonomous force takes on a new coloration when one restricts the term "technology" to those systems and assemblages that embody standards. On the one hand, evolutionary anthropology, going back to the nineteenth century evolutionists, states that technology was the driving force in history: like fire delivered by Prometheus, technology was a gift from the gods of Science that propelled human progress. On the other hand, more recent interpretations associated with the Social Construction of Technology is that technologies (and other instrumentalities) evolve through creative recombinations of artifacts, problems, and relevant social groups (Bijker et al., 1987). Both sides of this debate are represented in recent work on technological evolution.

The resolution offered here is first to restrict the concept "technology" to those instrumentalities that embody autonomous representations—standards, but also promotional documents—(usually associated with civil and state projects), and second to observe that within the developmental spaces defined by these standards, technological choices *are* relatively immune to social pressures, and technology *is* an autonomous force. More precisely, technological choice becomes a privilege available to a select few, an élite that has been socialized into its values and its epistemology.

Among the numerous representations of technology, including promotional, planning, and maintenance documents, standards occupy a privileged status. As described by Lampland and Star (2009), standards, including engineering standards, are embedded within communities of practice[9] (Lave & Wenger, 1991; Wenger, 1999) "codify, embody, or prescribe ethics and values," and increasingly constitute a worldwide, logically integrated system (2009, p. 5). Standards are coercive representations of large-scale systems: Developing a computer system based on Sumerian numerical notation (see Chrisomalis, 2010 for a description) would require only a small effort of actual coding, but a large, and probably ultimately frustrating effort, complete with PowerPoint™ presentations and numerous memos and written reports and task force meetings, to persuade standards bodies such as ISO[10] to extend the ASCII[11] character set. (Acronyms are used here to hint at the social boundaries that define these discourse régimes.) Standards are hegemonic régimes whose coercive character is felt only as they are extended beyond their original community of practice: In the Republic of China, for just one example, adoption of Western standards such as ISO9000[12] is most typically announced with a Roman orthography, making it clear both that the factory is "up to date" and that this is a foreign discourse.

Standards documents by themselves tend to be dry, lifeless tracts. The other autonomous representation of a technology, particularly those that have not yet been field tested, are promotional documents and displays, including museum exhibits, which grew in popularity in Victorian England. In this regard there is a rich history from the nineteenth and twentieth century of defining and promoting the acceptance of the technological devices that ultimately created the modern world.

A fundamental issue in the invention and innovation developmental cycle of any technology, as it emerges from the laboratory and is initially used by wider social circles, is *diffusion*[13] (Rogers, 2003), the process of early adoption and field testing, proving the system in a social context beyond that of its original development. Diffusion, a process of "enrollment," is central to the social construction of any technology. In software, "beta" refers to a version that is not yet fully mature, yet distributed with the expectation that users will report any unexpected problems. The labors of "early adopters" such as these are one of the externalities of technological development. Unless early adopters have some expectation that the technology will deliver benefit

[9] Anthropologist Jean Lave and computer scientist Eitenne Wenger developed the concepts of Situated Learning and later, Communities of Practice based on a theory that learning is essentially a social process.

[10] International Standards Organization.

[11] American Standard Code for Information Interchange.

[12] ISO9000 is a quality standard adopted by the International Standards Organization. For most supply chains in the West, its adoption is mandatory.

[13] In its 5th edition, Everett Rogers' *Diffusion of Innovations* (2003) documents in detail the full innovation process from the inception of an idea to its adoption by various social groups which are described as "adopter categories." Each adopter category has distinct characteristics that must be taken into consideration when introducing a new thing or idea.

that compensates for their efforts and inconvenience, or some other motivation, they will not adopt. The emerging literature of "disruptive innovation" suggests that these benefits are not always described as "improvement" (Christensen and Raynor, 2003). Early adopters *have* been persuaded to experiment and adopt by an expectation that technology would in some manner improve their lives, their communities, and their world. This assumption of inevitable technological progress is in some manner an article of faith in the industrial world, especially in the United States.

In short, representations—standards, documentation, museum displays, promotional tracts—are intrinsic to the technology of industrial societies and not simply superficial decoration. From these representations and accounts have emerged a set of technological aesthetics, combining images having positive appeal to certain groups, with a structured sense of what technology *ought* to look like. The most notable of these, of course, is the "High Tech" aesthetic, a set of images and styles that convey futuristic possibility. Rooted in the modernism of Raymond Loewy and Norman Bel Geddes, and the rise of industrial design, High Tech, with its emphasis on smooth surfaces and clean edges, often with a single focal point that conveys power, motion, or functionality, in contrast to, say, representations of authority, harmony, or sociality in medieval or classical aesthetics.

High Tech, we should emphasize, is an aesthetic concept at best loosely related to engineering issues of functionality, systems design and integration, or design tradeoffs. The term is more promotional than constructive or analytic. Although a critical analysis of High Tech is waiting to be written[14]—usage of the term took off like many other exciting innovations in the 1980s—we can observe for the moment that it is an assertion of technological authority contemporaneous with discussions of globalization, the breakdown of earlier state-based forms of authority, and the rise in experimentations with identity. The High Tech aesthetic has been adopted by numerous types of designed artifacts, including clothing, furniture, and personal accessories, whose engineered content is, at best, modest.

If High Tech is the official aesthetic of a technological society, then dissident voices are provided by at least two other aesthetics, Harley-Davidson, and assorted science fiction genres including cyberpunk and steampunk. We are adopting the name of a motorcycle to characterize an entire countercultural set of images that combine horsepower, working class identity, masculinity, and defiance of the state. This aesthetic burst into public consciousness with the 1953 movie, "The Wild One," in which Marlon Brando epitomized young, rootless men, attached only to each other and to their motorcycles. The Harley-Davidson and similar motorcycles, in contrast both to the Vespa and other European motor-scooters and to the sleek lines of automobiles, proudly display in their angularity their power, muscularity, and masculinity.

Science fiction, by contrast, presents a future of technological utopias and dystopias, where magical, energetic devices either embody the coercive authority of the state and other institutions or are used as tools of rebellion against it. Early science

[14] Various publications have been trending in this direction most recently *Burn Book* by Kara Swisher (2024) which documents the author's decades long career as a technology journalist.

fiction, perhaps exemplified by Jules Verne, whom we discuss in multiple locations, featured fantastic capabilities, including voyages to the moon and visitors from Mars, with the state, at best, playing a bit part. A more recent variant on this, cyberpunk, features the interactions of marginal characters, including computer hackers and street gangs, with cutting edge information devices and capabilities. Perhaps best exemplified by William Gibson's novel *Neuromancer* (1984) , cyberpunk is a countercultural riposte to a technological society.[15]

Technological celebration, science fiction, cyberpunk, steampunk, and related genres create the conceptual spaces within which technological capabilities can be imagined before they are invented. Michel Nadar landed on the moon in 1865, more than a hundred years ahead of astronauts Neil Armstrong and Buzz Aldrin, in Jules Verne's *De la Terre à la Lune* (Verne, 1865). The fact that Verne's transport device—a projectile gun—ultimately failed in field tests did not prevent his fictional treatment from inspiring millions of technological dreamers over the next century, or a 1998 Hollywood celebration of the Apollo program, *From the Earth to the Moon*, a TV mini-series starring Tom Hanks.[16] In contrast to official celebrations of technology, whether in promotional literature or in futuristic monuments, many of these genres interrogate the relationship between technology and the state and powerful institutions in the corporate world.

Finally, we might consider the austere conventions of technocratic rationality as having their own aesthetic allure. Technocratic rationality, with its reliance on quantification and its supposed immunity to social prejudice, projects an air of efficiency that does not just beg but actually forecloses questions of the ends toward which that efficiency is deployed. As a dominant form of legitimation in late modernist societies, technocratic rationality is only beginning to be interrogated from a critical perspective (See Riles, 2004 or Merry, 2011, or Roszak, 1969).

The rise of these genres is contemporaneous, roughly, with the scaling up of industries and functions, whether transportation in the nineteenth and early twentieth centuries, communication in the mid-twentieth century. Contemporaneous with this has been the displacement of innovation away from the workshop of the self-educated inventor such as Edison into government and corporate laboratories staffed by scientists and engineers with advanced academic degrees (Friedel, 2007:319ff.; Israel, 1992).[17] The contemporaneity and convergence of these trends—increasing scale, autonomous representation, and the professionalization of functions that were

[15] A variant on cyberpunk, "steampunk" takes the basic cyberpunk conventions and places them in a nineteenth century world of steam power and clattering mechanical devices, beginning (but certainly not limited to) Charles Babbage's "analytical engine," a forerunner of the modern computer.

[16] https://www.google.com/search?client=firefox-B-1-e&q=From+the+Earth+to+the+Moon%2C+a+TV+mini-series+starring+Tom+Hanks.

[17] Other important trends were the professionalization of business. Contributing factors include the Industrial Revolution, the emergence of business schools, the development of management theory, the rise of professional associations, the codification of ethics and standards, and the growth of corporate structures. It was in this period design was also becoming a professional occupation. The Industrial Revolution played a role here as well, plus other factors such as the Arts and Crafts movement, the founding of design schools, and the rise of corporate and industrial design.

once considered rudely mechanical—amounts to a national technological turn, a new *conscience collective* in which scale, power, and coercion are celebrated. The hand of the state—visible and order-creating within the standards that create spaces for technological development, invisible and celebratory in the promotional literature and displays, yet evil and visible in cyberpunk—is the back story of all technological development.

In sum, the construction of "technology" and the construction of the modern world are roughly coincident, both as contemporaneous events and mutually reinforcing trends. Our intent in the balance of this volume is to explore these dynamics in order to construct an imagination of a world beyond the modern construction of hegemonic technology.

"Technology" as a Key Word

The conclusion of all this is that technology is a *key word,* a term that knits together disparate concepts into a unified whole. Key words, in the view of Raymond Williams (1976), include "family," "fiction," and "folk" as well as more than a hundred others in Williams' *Vocabulary of Culture and Society.* A key word metonymically embraces contradictions and inconsistent meanings, thus bringing together the complexity of contemporary discourse into a unified whole. Other examples of key words include constructs such as nation, tribe, and factory.

As we noted, this key word, technology, did not exist before 1821, when Bigelow's *Principles of Technology* was published. The uniting of *techné* (skill or craft) with *logos* (the authority of the written word) started in the third decade of the nineteenth century, with the growth of institutions such as the Massachusetts Institute of Technology (founded in 1861 and subsequently followed by numerous other Institutes of Technology). Understanding technology as a key word unlocks several mysteries that we have been contending with. At the top of that list is the multiple and inconsistent meanings of the term technology.

The key word "technology" embraces multiple meanings. Foremost, of course, are tools and instrumentality, but also important is the totemic character of technology, the manner in which it communicates the user's identity, as described in the previous chapter. Also important, yet less noted, is that technologies embrace *totalities*, all-encompassing views of the world and society. The automobile in America, for but one of many examples, embraces not simply transportation but also identity, sociability, and community.

In sum, "technology" as a concept (contrasted to premodern toolkits) is very much a modern, western construction, affording a dynamism to modern, Western society that earlier terms and tools lacked. As its dynamism and tensions (and disruption) become apparent, we can begin to look beyond the modern construction of *hegemonic technology*, the technological systems or practices that have become dominant, pervasive, and widely accepted within modern society, shaping social norms, cultural practices, economic structures, and political power dynamics.

The construction of technology and the construction of the modern world are mutually reinforcing processes: not simply one-way technological determinism, the idea that "technology drives history," but also a reverse process, the idea that modern history drives technology, considered here not just as toolkits, but rather as an ensemble of artifacts, expectations, and problem-solving. The architects of the technological society, the engineers who built not only the devices but the entire edifice of expectations and institutions that fulfill our dreams, are the subject of our next chapter.

References

Anderson, B. (1983). *Imagined communities: Reflections on the origin and spread of nationalism.* Verso.
Berman, M. (1988). *All that is solid melts into air: The experience of modernity.* Penguin.
Bijker, W. E., Hughes, T. P., & Pinch, T. (Eds.). (1987). *The social construction of technological systems: New directions in the sociology and history of technology.* MIT Press.
Boorstin, D. J. (1972). *The image: A guide to pseudo-events in America.* Atheneum.
Bury, J. B. (1921). *The Idea of progress.* London: Macmillan and company.
Bury, J. B. (1932). *The Idea of progress: An inquiry into its origin and growth.* New York: Macmillan.
Christensen, C. M., & Raynor, M. E. (2003). *The innovator's solution: Creating and sustaining successful growth.* Harvard Business Review Press.
Chrisomalis, S. (2010). *Numerical notation: A comparative history.* Cambridge University Press.
Dumont, L. (1970). Caste, racism, and stratification. In A. Tuden & L. Plotnicov (Eds.), *Social stratification in Africa* (pp. 102–109). Free Press.
Ellul, J. (1964). *Technological Society.* A.A. Knopf. (Original work published 1954).
Emerson, R. W. (1847). *Ode, inscribed to William H. Channing.* In *Poems* (pp. 166–169). Phillips, Sampson and Company.
Friedel, R. (2007). *A culture of improvement: Technology and the Western millennium.* MIT Press.
Gibson, W. (1984). *Neuromancer.* Ace Books.
Hughes, T. P. (1983). *Networks of power: Electrification in Western society, 1880–1930.* Johns Hopkins University Press.
Israel, P. (1992). *From machine shop to industrial laboratory: Telegraphy and the changing context of American invention, 1830–1920.* Johns Hopkins University Press.
Jacobsen, A. (2013). The Needham question updated: A new take on Joseph Needham's grand question in the light of the influence from Chinese philosophy and values on science and technology in China. *East Asian Science, Technology and Society: An International Journal, 7*(2), 127–151. https://doi.org/10.1215/18752160-2159451
Lave, J., & Wenger, E. (1991). Situated Learning: Legitimate Peripheral Participation. In: Cambridge University Press.
Lilienthal, D. E., Eaton, J. W., Pearson, F. A., & Paarlberg, D. (1945). TVA—democracy on the march. *Science and Society, 9*(1), 94–96.
Lin, J. Y. (1995). The Needham puzzle: Why the Industrial Revolution did not originate in China. *Economic Development and Cultural Change, 43*(2), 269–292.
Marx, K. (1904). *A contribution to the critique of political economy* (N. I. Stone, Trans.). The International Library Publishing Co. (Original work published 1859).
Merry, S. E. (2011). Measuring the world: Indicators, human rights, and global governance. *Current Anthropology, 52*(S3), S83–S95. https://doi.org/10.1086/657241
Needham, J. (1954). *Science and civilization in China* (Vol. 1). Cambridge University Press.

Pfaffenberger, B. (1992). Technological dramas. *Science, Technology, and Human Values, 17*(3), 282–312.
Riles, A. (2004). Real time: Unwinding technocratic and anthropological knowledge. *American Ethnologist, 31*(3), 392–405. https://doi.org/10.1525/ae.2004.31.3.392
Rogers, E. (2003). *Diffusion of Innovations* (5 ed.). Free Press.
Roszak, T. (1969). *The making of a counter culture: Reflections on the technocratic society and its youthful opposition.* Doubleday.
Turnbull, S. (1993). *The book of medieval wisdom: A treasury of proverbs and quotations.* Blandford Press.
Verne, J. (1865). *De la Terre à la lune.* France: Pierre-Jules Hetzel.
Wenger, E. (1999). *Communities of practice: Learning, meaning, and identity* (1st ed.). Cambridge University Press.
Winner, L. (1986). *The whale and the reactor: A search for limits in an age of high technology.* University of Chicago Press.
Wolf, E. R. (1982). *Europe and the people without history.* University of California Press.

Chapter 4
An Engineer's Perspective

Abstract In this chapter, we establish the boundaries of an engineering approach to technology, considering technology from an engineering perspective and stressing the importance of standards and other forms of autonomous representations for defining technology. We explain *autonomous representations* as descriptions of the technology that are independent of the actual tool or object or implement yet are at the core of technology as it is understood by engineers. Finally, we examine how the leading authorities of the technological society, the engineers, have carved out a distinctive role in contemporary society. This role or authoritative voice shapes many of our expectations in the technological and post-technological society.

While conceptualizing this book, one of the authors (Batteau) worked for seven years as a systems engineer, analyzing, designing, and building computer-assisted logistics systems for the Air Force and private contractors. In the course of this work, he became acquainted not only with the technical details of systems engineering, but also with the professional culture of material and systems engineers, spanning both private and public sectors. Foremost among the traits of these cultures was a pragmatic outlook, "Does it work?" with less questioning of the ultimate goals of "it." The tools and techniques that he used, including functional and data modeling and simulations, provided a workable *representation* of the systems which he was analyzing and building. These models and simulations were not simply markings on a sketchpad, but rather *formalized* descriptions of the systems, their functionality, and their flows of information. As we discussed in the previous chapter, *representation* in addition to assembling is a key part of engineering practice.

Engineers are among the central figures of the technological society. The engineering profession, which is as old as civilization (and older than many other *professions* including the law and medicine), is central to the culture of cities and empires. Engineering was neglected because it was not joined to *logos,* the authority of the written word. Only with the Industrial Revolution and the creation of the first technological institutes such as the Massachusetts Institute of Technology did engineering become a respectable *profession* with an institutional foundation comparable to other contemporary institutions.

A. Batteau and C. Z. Miller, *Tools, Totems, and Totalities*,
https://doi.org/10.1007/978-981-97-8708-1_4

Engineers do not just create new technologies. They also build, maintain, and improve the structures and institutions of civilization, some of which have lasted for thousands of years. For most of history this was uncelebrated until the creation of *technology.* There was little connection between *techne* and *logos*, the authority of the written word.

In this chapter we examine how the leading authorities of the technological society, the engineers, have carved out a distinctive role in contemporary society. This role or authoritative voice shapes many of our expectations in the technological and post-technological society.

The Ancient Engineers

The use of a formal syntax to describe form and functionality goes back to the earliest years of engineering, to the construction of the Egyptian pyramids and the Babylonian streets. In fact, we can examine all the ancient engineering wonders in terms of form, function, and flow of information and energy, whether for dynamos and water wheels of rivers, or the traffic in a city.

How the formalities of engineering evolved from the mysteries of craft guilds into the standards of national systems is the major story we want to tell here. Some of the ancient engineers included Vitruvius (first century BCE) and Marcus Vipsanius Agrippa, mentioned below, from the same century. One of the best remembered contributions of Vitruvius was the three orders, *firmitas, utilitas,* and *venustas*, which have endured as basic principles of architecture for two millennia. *Firmitas*, or solidity or strength, is the building's ability to withstand the elements of nature for years or centuries, in contrast to durability if mud huts or yurts. Over time architects have developed the tools to calculate solidity and strength, so that they can design structures that endured as long as they could image, for millennia, actually. *Utilitas,* or usefulness, is the matching of a structure to its intended purpose, whether a dwelling or a workshop or a temple, whether it contains domestic quarters or soaring basilica. *Venustas*, or beauty, is the aesthetic aspect of the building and is a notable aspect of all architecture since ancient civilizations (Vitruvius, 1914).

Another notable Roman architect was Apollodorus of Damascus, who lived in the first century CE and who notably designed Trajan's column and the Forum in Rome (Fig. 4.1).

Apollodorus was a military engineer who designed several bridges across the Danube to enable Roman armies to extend into the provinces. Also notable was Marcus Vipsanius Agrippa, the architect who designed the Pont du Gard, an aqueduct in southern Gaul in in the first century CE. Vipsanius Agrippa, who was born a plebian (commoner), distinguished himself not only by his engineering talents, notably by designing the Pantheon, but also by his relationship to his father-in-law the Emperor Augustus (Fig. 4.2).

Throughout history, engineers have been building the structures that define civilizations, whether roads or palaces or the Egyptian pyramids. Many of these have

Fig. 4.1 Trajan's column (c. 113 AD)

survived since antiquity: the Mayan pyramids, for example, from 1000 years BCE are perhaps the oldest structures in the Americas. Similarly, the Aztecs established a civilization from the fourteenth to the sixteenth centuries, only to be decimated with the arrival of Europeans. Likewise, the Roman Coliseum, constructed in the first century CE, still stands as a monument to Roman civilization. These monuments are not simply conveniences, but rather stand as eternal statements of the grandeur of their civilizations. They are as much aesthetic and political statements as they are engineering accomplishments (Fig. 4.3).

Fig. 4.2 Pont du Gard (c. 17 BCE)

Fig. 4.3 Mayan pyramid

In sum, engineering has always been central to civilization, notably that *civilized* societies, in contrast to tribal aggregations, have found or search for ways to live together peaceably. Living together peaceably entailed not simply avoiding warfare, but also building settlements, public squares, and places of worship to gather. This peaceful collective coexistence, along with proclaiming the emperor's grandeur, the reverence felt for the gods, and the facilitating of commerce, has been the overriding objective of engineering for millennia.

Human Factors Go Global

Antiquity had notable engineering achievements, many of which either in their legacy or in their actual constructions survive even today. They have been able to survive because they established traditions of engineering achievements that different civilizations, both Western and Eastern, have built upon. Eastern traditions such as Chinese architecture must grapple with the same laws of physics while incorporating aesthetic and ethical standards such as *Feng Shui*. Feng Shui (literally, "wind water") is an important cornerstone in Chinese architecture. It is still taught at Singapore Polytechnic and engineers, and architects use it in their practice. Feng Shui refers to the alignment of structures with patterns of wind and water to optimize the flow of cosmic energy, or *Qi*, with the structure. Qi is central to many Chinese technological and medical practices yet is a totally foreign concept in the West. As a consequence, Chinese engineering standards have a different emphasis from Western standards, not only in their aesthetics but also in basic structural principles. For example, in China there are standards for light-signaling devices, door locks, rear-view mirrors, child safety seats, and dozens of other building and automotive devices.

One example of Eastern culture affecting engineering design is what the aerospace engineer Hung-Sying Jing has identified as the "guanxi gradient" (Jing, 2006). Guanxi is a Mandarin word meaning (approximately) "connection," although its meaning is far deeper than that. To "have guanxi" means acquiring connections, one of the most important social resources in Chinese culture. This contrasts with the individualism of many Western cultures. The study of "human factors" in aviation, a discipline that emerged during World War II when it was discovered that more pilots' deaths were due to human error than due to enemy encounters. "Human Factors" includes ergonomics, the adaptation of machine design to the mean physiological capabilities of the human body. In the Western context, human factors include communication efficiency, monitoring, and clearly defined roles. A key discipline in aviation human factors is "crew resources management," which includes communication, monitoring, cross-checking, leadership, decision-making, and situational awareness. The pilot-in-command, typically in the left-hand seat of the cockpit, must articulate his actions to the copilot (communication), who reads back the instructions (cross-checking) to confirm that he has understood them. In a cross-cultural survey Jing found that *guanxi,* as a gradient capable of being measured, had a strong correlation with accident rates among Taiwanese pilots. This discovery (or rediscovery) that cultural variables are closely related to technological performance makes clear that technologies are intimately embedded within a variety of cultural contexts (Jing and Batteau, 2015).

The engineering of human factors attempts to fine-tune the human/machine interface, combining an awareness of communication and authority patterns with the technical requirements of the device (the aircraft). It has been notably successful: in proportion to passenger-miles traveled, air transport is the safest form of transportation available, far safer than automobiles. This success comes from the recognition that to optimize the man–machine interface both the machines (aircraft) and humans (flight crews) need to be tuned to each other. When the human context varies, notably from East to West or North to South, whether in communication patterns or expectations of authority, this fine-tuning must be adjusted.

An Institutional Perspective

Generalizing from this, we note that any engineered system can impact multiple social objectives, and the prioritizing of one objective or set of objectives over others is always an institutional and not a technological decision. Air transport has been optimized for safety, and efforts to optimize it for speed, such as the currently-defunct Supersonic Transport (SST), or even for passenger comfort, have gone nowhere. In America, passenger cars and cityscapes are optimized for commuting, often at the expense of aesthetics. In Europe, by contrast, cityscapes are optimized for communality, and the proliferation of sidewalk cafes in Paris, and their almost complete absence in nearly every large American city testifies to the contrasting institutional priorities of America and France. These institutional priorities include an emphasis on private goods and Gross Domestic Product (GDP) in the United States in contrast to sociality, neighborliness, and the enjoyment of life. The recent controversy in France regarding extending the retirement age from 62 to 64 is not simply a controversy over actuarial possibilities, but more importantly a controversy over what it *means* to be a Frenchman (Fig. 4.4).

Jing found that "human factors" has additional dimensions when considered in a cross-cultural context. Engineering standards, whether the voltage of power grids or

Fig. 4.4 Paris sidewalk cafe

the design of machines, are always *national* projects, and different nations or clusters of nations have developed their own standards. For example, in China, approximately 60 standards exist for such matters as food, building construction, and roads. Chinese standards include "Descriptive rules for bibliographic references" and road traffic signs and markings. These standards have the force of law and are essential for any engineering project in China.

We can see that engineering, in addition to its technical details of mass and energy and information flow and storage, contains an *institutional* edifice, that is, a set of social relationships and compromises, wrapped in a cloak of scientific and state authority, that ensures both its durability down through the ages and its harmony with other aesthetic, ethical, and legal cultural norms. Building this institutional edifice, a work of more than two millennia is no less a foundation of the technological society than the girders and trusses that uphold architectural monuments such as the Coliseum and the Pont du Gard.

What has changed since the Industrial Revolution is the recognition that engineering, along with the useful arts in general, assumed a leading role in society, alongside the liberal arts and governing arts. The technological society which this created (Ellul, 1964) is defined now by its engineering accomplishments, even as other institutional accomplishments whether in government or worship or education or aesthetic communality are slighted. In the technological society, all is focused on technique, on the means to the end, and the primary criterion for judging technique is efficiency. Thus, commuting alternatives are judged primarily by time and distance, with perhaps safety factored in as well. Each of these components can be easily measured. What is less easily measured are such imponderables as comfort, neighborliness, landscape vistas, enjoyment of the ride, impact on the environment, and work/life balance. All of these are of varying importance and can be measured, but with greater difficulty than time and distance and energy. Consequently, a focus on efficiency and technique reverts to the most easily measured variables, possibly at the expense of less tangible variables such as aesthetics or community solidarity.

The ethos of engineering entered management through Frederick Taylor's "scientific management" which used time-and-motion studies and stopwatches to reduce skilled craftsmen to appendages of the machine. The machines that Taylor studied whether on an assembly line or in a machine shop were strictly subject to the Newtonian laws of motion. Taylor calculated how much time was required for a worker to move from one machine or one task to another and optimized the use of the workers (hands) to fit the motion of the machine. These then became standards for man–machine interaction. The workers thus became appendages to the machine, and skilled craftsmen were reduced to unskilled machine tenders. Although this increased the overall productivity of the machine, it was at the expense of the quality of the work and the self-respect of the workers. The entire discipline of industrial engineering, which took off during World War II, a war that was won by American industry and the "arsenal of democracy," was outproducing the Axis powers in weapons, tanks, aircraft, and ships. It arguably made technology the handmaiden to imperial ambitions.

One outgrowth of Taylor's scientific management was the reaction by Elton Mayo in the "Hawthorne studies"[1]. Attempting to optimize the man/machine interface, Mayo and his colleagues manipulated multiple variables, including background music, lighting, and air conditioning. Much to their surprise they found that practically anything improved productivity, leading to two conclusions: first, that just as important as the man–machine interface was the man–man interface, the opportunity to bond with one's coworkers. The second conclusion was that the subordinate–supervisor interface was an important part of productivity. The realization that the boss was paying attention was perhaps more important than any sense that the machine was paying attention. Mayo's findings are summarized in Bell's classic "Adjusting Men to Machines" (1947).

Heroic Engineering and Its Consequences

The "arsenal of democracy," Franklin Roosevelt's characterization of Detroit at the height of World War II, combined the manufacturing capability pioneered by Henry Ford and others with a singular commitment to building the tanks, guns, trucks, and aircraft that won the war. During World War II, aircraft production in the United States ramped up from fewer than 10,000 per year in the years before Pearl Harbor to 96,000 in 1944; by contrast, in Germany went from 10,000 to 39,807 in the same years, and in Japan from 4768 to 28,180. In hindsight, the Allied victory was inevitable. The combination of industrial capability and unified national purpose has perhaps no historical parallel and might even be seen as having contemporary relevance less in America's technological agenda and more in terms of finding ways to rebuild national unity. Over time America has prioritized economic growth and technological capability not just to the neglect, but at the expense of national unity.

The current breakdown in national unity, which many have commented on, has many roots: increasing inequality, racism, and the lingering legacy of slavery, and the fact that, unlike in 1941, the existential threats are ill-defined. The fact that some of the leading technologies of today, notably Facebook and Twitter, are optimized for spreading emotion-arousing memes or "clickbait" is, we would argue, central to the breakdown of American unity. Tribalism, a splitting of the world into competing factions, is a consequence of contemporary technology.

Extrapolating from this beyond technologies such as social media, we can observe many instances where technologies reinforce social separations. Urban design, for example, is optimized for the automobile, yet the price of this efficiency is a breakdown into neighborhood enclaves and increasing segregation of institutions. Transportation systems including automobile and airplanes create separations rather than community, and information systems all too frequently sow misinformation.

---

[1] The Hawthorne Studies (Muldoon, 2017) and the findings of Mayo and his colleagues have been the subject of much discussion and debate. https://www.emerald.com/insight/content/doi/10.1108/JMH-09-2016-0052/full/html.

In the technological society, *everything* is an appendage of the machine, and norms of mechanism, notably efficiency, a lack of friction, and transparency of input to output govern all. The consequence of this for other social values, whether collegiality or fraternity or worship, is profound; the best that the technological society has been able to offer is the "religion of technology" (Noble, 1997).

In David Noble's *The Religion of Technology* (1997), certain contemporary systems, including nuclear weapons, space exploration, artificial intelligence, and genetic engineering have acquired almost theological status. Most if not all of these are solutions in search of problems. Nuclear weapons, which ended World War II, have not been used since except as deterrents. Space exploration, other than earth-orbiting satellites, have no direct productive value and can be sustained only with massive cash infusions. The idea of mining anything on Mars and bringing it back to Earth in a cost-effective way remains off in some distant future. According to Noble, the "religion of technology" offers transcendent values—omniscience, all-consuming power, immortality—that contemporary Christianity (and possibly other, non-European religions) have failed to deliver. Until the relatively recent emergence of a fundamentalist, white nationalist version in the US, Christianity had lost much of its sacred aura amid corruption and scandal in multiple denominations. Technology steps into this void.

An important part of the human social experience is spirituality and religious worship, whether in a church, synagogue, or mosque, and efforts to make this more efficient, whether through megachurches—suburban congregations embracing thousands of worshippers—or live-streaming televangelists over the Internet, all degrade the experience. *Reverence*, the communal experience with fellow congregants in the same room, is not the same when it is online or over the air. While congregants and pastors may go through the motions of worship, there is clearly not the commitment or sacrifice of in-person fellowship. In fact, the dynamic of *sacrifice*—the voluntary surrender of value—while bestowing sacred value on the sacrifier is antithetical to the efficiency promised by technology. Most, perhaps all religions have sacrifice at their core, yet the religion of technology promises only ease and efficiency.

It is not difficult to demonstrate that efficiency can often be *inefficient*, less in terms of mechanical inputs and outputs and more in terms of overall social objectives. A myriad of examples abound: the design of city streets, optimized for automobile, comes at the expense of green spaces and footpaths; mass transit systems, whether subways or buses, optimize the flow of workers and shoppers, yet at the expense of both pollution and quality of the transit experience. Urban density, perhaps an efficient use of real estate square footage, after a point degrades the urban experience. The contrast between cities of Europe and America is adequate testimony to this. The shadow cast by skyscrapers is now subject to zoning regulations, and "sunlight rights" are in some cities an important part of the calculation of real estate values. "Sunlight rights" and other externalities such as pollution and traffic congestion are epicyclic refinements on the inefficiencies of contemporary urban design. Lutz and Fernandez's Carjacked (2010) explores how American society has been adapted to Henry Ford's horseless carriage.

Similarly, factory production, whether of automobiles or clothing, is optimized for the ratio of output and expense over worker comfort and externalities such as pollution. Efforts to rebalance this include both the industrial union movement of the mid-twentieth century and the environmental movement of the 1970s onward. Today a movement for "work/life balance" (Gragnano et al., 2020) pushes back against the machine.

The calculus of efficiency works *only* when one focuses on a limited range of functionality, whether traffic movement or residential comfort, and when one clearly specifies the context of the functionality. The focus of efficiency often devolves to a least-common-denominator focus on variables that are more easily quantified. Shopping malls, for one of many examples, are very efficient for facilitating consumption, but very *inefficient* for facilitating neighborliness. Consumer purchases can easily be quantified, but neighborliness cannot. City streets are very efficient for facilitating traffic movement between suburban neighborhoods and workplaces, but very *inefficient* for bringing relatives and neighbors together. "Neighborliness," while measurable, is more difficult to quantify than traffic flow and thus is slighted at the expense of the more quantifiable variable.

After World War II, America's notable contribution to architecture was the creation of the shopping mall, a "third place" between home and work and which quickly replaced the public square as a locus of conviviality. Although the definitive architectural history of the shopping mall remains to be written, we can note for the moment that the rise of the shopping mall was congruent with an increased emphasis on the consumption of private goods, often at the expense of both public goods and club goods such as neighborhood solidarity.

Viewing the technological society from an engineer's perspective reveals numerous *inefficiencies* that go unremarked. For example, skyscrapers that assert an imperial dominance, whether the Empire State Building or the World Trade Center, while (possibly) an efficient use of urban square footage, notably are inefficient in their use of commuters' time and other externalities such as the cityscape and aesthetic considerations. A cityscape that is optimized for capital efficiency is demonstrably sub-optimal for neighborly gatherings.

We must also note that engineering values are *always* embedded within an institutional context, whether assertions of imperial grandeur, the defense of freedom, or the all-seeing eyes of the state. China's "social credit" system, described in the next section, is one perhaps extreme example of the assertion of hegemony and imperial values through a technological edifice.

In sum, the engineers' perspective, focusing on variables that are quantifiable, neglects the aspects such as neighborliness and sociability that are not so easily measured.

Cosmology and Corruption

Institutions always embrace cosmological values of "Who we are" and "Where do we fit into the universe?" Empires that bestride the narrow world have very different cosmology from tribal villages, and erect roads and bridges—monuments, actually—not just for moving people around but for proclaiming imperial grandeur. In architecture this is obvious with grand buildings making an imperial statement. Mansions, likewise, proclaim the comfort and ease of the lord of the manor, even if they are in suburbs. On September 11, 2001, the Statue of Liberty was *not* attacked because its statement regarding "huddled masses yearning to breathe free" echoed a sentiment that has been felt around the world, down throughout the ages; by contrast, the twin towers of the World Trade Center, like the Pentagon, were assertions of hegemonic imperial might and directly targeted.

Cosmology is a taken-for-granted part of engineering practice because like the rising of the sun every morning it seems normal, "natural." This is practical because most technologies operate only within a singular cultural context, whether the national culture of an infrastructure or the professional culture of an occupational specialty. It is only when technologies leap these boundaries, whether air transport or power grids, that these taken-for-granted assumptions must be examined and renegotiated. Power grids must contend with different national standards for electrical current and even different regional standards for cooperating with the neighbors, and air transport must renegotiate its safety standards as it moves from center to periphery. With power grids, the difference among multiple nations over voltage, cycles-per-second, and even alternating versus direct current, as well as local attitudes, means that the building on a "grid"—a common pool resource for sharing generating capacity and adjusting peak loads among neighbors—must be socially and politically negotiated. The "grid" is technologically constrained but *not* technologically determined, and the decision to participate or not participate in the grid reflects local attitudes: In 2021, Texas had previously chosen not to participate in the Western Interconnection (a grid covering most of the states west of the Mississippi), a weather event led to a blackout for several days that was confined to Texas, as we discussed in Chap. 2. Excepting Texas, the grid has proven to be remarkably durable at guaranteeing power despite weather events, and Bakke (2017) has documented the numerous compromises that have made it work. Similarly, with air transport, as one moves from core nations (primarily Europe, North America, and Asia) to the periphery (notably Africa and Latin America) communication protocols must be adjusted to assure flight safety, in terms of cross-checking and turn-taking, essential parts of discourse.

In fact, the balance of technological development on a global scale assures an *imbalance* between core and periphery. The *geometric* acceleration of technological development in core nations and regions, first observed by Intel founder Gordon Moore and later named "Moore's Law" (Moore, 1965; Schaller, 1997), when juxtaposed with the *arithmetic* pace of technological diffusion from core to periphery, suggests that this imbalance is inherent in the character of technology (Batteau,

2010). Large-scale technologies, whether air transport or the Internet, always create imbalances between core and periphery.

One part of any cosmology is its time horizon, and different cultures have different time horizons, both into the past and into the future. In America, the dismissive "that's history" contrasts with the affirmation in other parts of the world that "history" is very important. In France, the assertion that "c'est histoire" is a statement that "that is very important." The well-known short-term thinking of American management is actually repeated across numerous institutions and professions. For but one of many examples, infrastructure—roads, bridges, city streets—is in America engineered for a 50-year life span, and the well-known neglect of infrastructure in America is a familiar story. The American Society of Civil Engineering (ASCE) has documented in its "Report Card on America's Infrastructure" that over the past twenty years America has had below-average infrastructure. By contrast, Roman roads and bridges have lasted for thousands of years, even if they are no longer considered adequate for modern traffic.

Roads can be *arteries*, facilitating the flow of traffic; they can also be *monuments*, proclaiming the majesty of the state or the empire. The first roads in America were the post roads, connecting cities for the postal exchanges, such as the Albany/New York Post Road, and the provision in Article I, Sect. 8 of the Constitution that gave the Congress the authority "to establish post offices and post roads," thus making continental integration a national priority.

The cosmological horizons extend not just in time and space, but also into the depths of personal lives. In the 1990s China launched the "social credit system," a big data platform using facial recognition, location-tracking, and cell phone apps and other data points to track the movement of its citizens. The "social credit system" is not a great engineering breakthrough so much as it is an institutional breakthrough, extending surveillance capabilities in a society lacking privacy protections. Using advanced technology including big data, the social credit system extended the all-seeing eyes of the state, a capability which the Soviet Union could only imagine in the era before the Internet. "Big data," a concept that scarcely existed before the current century, like "cloud computing" (the storage of megabytes and terabytes of data on server farms located almost anywhere) reproduces the all-seeing eyes and all-knowing capability of Big Brother, now all over the world.

In conclusion, we can see that technological capabilities always exist within institutional and cultural contexts, and efforts to slip those surly bonds either end badly or require not so much fine-tuning as gross adjustment to conform to social expectations. The redesign of urban cityscapes to conform to the requirements of automobility, or the rearrangement of expectations of privacy in the age of cell phone apps and GPS capabilities, or the degradation of the craftsman's workshop into the efficient, Taylorized assembly line all speak to an adjustment of men to machines. While some may argue that overall, this has created greater affluence; in fact, the unequal distribution of (private) affluence and the decline of public goods and the commons in an "affluent" society, which John Kenneth Galbraith commented on more than 65 years ago, suggest less a Faustian bargain than a corrupt tradeoff. The institutional corruption that many have commented on, whether the decline of political parties into personality cults or the degradation of activities in the public sphere into grubby profit opportunities, from coffee-house discussions into shouting matches over social media, is less remarkable once one recognizes that technological capabilities, when let loose from institutional restraints, always corrupt the hands that hold them. As wondrous as these capabilities are in their own terms, in their own horizons of time and social space, of savings of time and energy (which, as we have seen, may be dubious), they are far more questionable when measured against the society's entire institutional apparatus.

The withering of institutional restraints in "the land of the free" is so obvious that it is rarely commented on. Institutions, whether in the form of national borders or public squares or public services, *always* provide the context for engineering achievements, and efforts to degrade these institutions in fact undermine technological achievements. Whether we consider the deregulation of air transport in the late twentieth century, or loosening zoning restrictions in many municipalities creating greater inequities in schools or public services, overall result in a deteriorating quality of the urban experience. The growing inequality and tribalism in America, which many have noted, is not simply due to the growing influence of neoliberalism and its emphasis on private accumulation at the expense of the public good, but also to the obsession with technological achievements. The "land of the free" is fated to crash into irrelevance.

This obsession can be seen in the celebration of technological systems such as Twitter that primarily sow discord, the reliance in some countries such as China on creating a new Big Brother (such as China's Social Credit system), fantasies such as colonizing Mars, or creating alternative realities (the "metaverse," described in Chapter 11), or most recently the get-rich-quick scam of cryptocurrency that supposedly frees money from institutional restraints (see chapter 11). The fact that some (perhaps many) are beginning to realize the dubious foundations of these suggests that we may be on the verge moving beyond technology.

Technologies aggregate energy, matter, and control, but most typically in the hands of those who own or control them. The increasing economic inequality of American society, which many have commented on, is paralleled and reinforced by an increasing technological inequality. The net result is a withering away of institutional restraints in the "land of the free" is an ironic consequence of the technological society: Freedom is the *opposite* of mutually agreed-upon institutional restraint.

As America focuses more on (corporate funded) technological achievements—faster, higher, farther, cheaper—it degrades the spirit of community which provides an unquantified counterweight to the fracturing of society. Any system that is optimized for least-common-denominator variables is an impoverished system.

In fact, one might venture that the institutional corruption of contemporary society is precisely a consequence of the society's enthrallment with technology, whether in its worship of context-specific efficiencies or its neglect of broader social objectives. This enthrallment with technology, which Lewis Mumford (1971) termed the "myth of the machine," which we examine in Chapter 7, after giving attention in chapters 5 and 6 to the role of design in creating a new society.

References

Bakke, G. (2017). The grid: The fraying wires between Americans and our energy future. Bloomsbury.

Batteau, A. (2010). Technological peripheralization. *Science, Technology, and Human Values, 35*(4), 554–574.

Bell, D. (1947). Adjusting Men to Machines. *Commentary, 4*, 79.

Ellul, J. (1964). *The technological society*. English translation of La Technique ou l'enjeu du siècle. New York: A. A. Knopf.

Gragnano, A., Simbula, S., & Miglioretti, M. (2020). Work–Life Balance: Weighing the Importance of Work–Family and Work–Health Balance. *The International Journal of Environmental and Public Health, 17*(3).

Jing, H.-S., & Batteau, A. (2015). *The dragon in the cockpit. How western aviation concepts conflict with Chinese value systems*. Burlington, Vermont: Ashgate Publishing Co.

Lutz, C., & Fernandez, A. L. (2010). *Carjacked: The culture of the automobile and its effect on our lives: The culture of the automobile and its effect on our lives*. St. Martin's Press.

Moore, G. E. (1965). Cramming more components onto integrated circuits. *Electronics, 38*(8), 114–117.

Muldoon, J. (2017). The Hawthorne studies: An analysis of critical perspectives, 1936–1958. *Journal of Management History, 23*(1), 74–94. https://doi.org/10.1108/JMH-09-2016-0052

Mumford, L. (1971). *Myth of the Machine: Technics and Human Development*. Mariner Books.

Noble, D. (1997). *The religion of technology: The divinity of man and the spirit of invention*. Knopf.

Schaller, R. R. (1997). Moore's law: Past, present and future. *IEEE Spectrum, 34*(6), 52–59. https://doi.org/10.1109/6.591665

Vitruvius, Pollio. (1914). *The ten books of architecture*. (M. H. Morgan, Trans.). Cambridge: Harvard University Press.

Chapter 5
The Design Perspective

Design as inquiry is a way of engaging the world. It seeks to understand the relationship between things, cultural frames, action/behaviors, psychological effects, and contexts. It seeks to both understand and make meaning. And it sees opportunity for the making of meaningful things that can influence the frames, actions/behaviors, psychological effects, and contexts in which they are embedded.
Design Unbound, Pendleton-Jullian and Seely Brown (2018, p. 12).

Abstract In this chapter, we examine technology from a design perspective, noting how the aesthetic dimensions of technology, notably modernism, figure into popular understandings of technology described the in previous chapters. For most of history, designers and artists occupied separate social and semantic spaces from rude mechanicals and other artisans. With the advent of modernism in the nineteenth and twentieth centuries, design became a critical element of technology and a critical *instrument* in shaping society and its built environment. The purpose of this chapter is to sketch the contours of design from a perspective that is beyond technological hegemony in which technology is not perceived as the best and final answer to every need or problem. This perspective takes a critical and nuanced view of technological hegemony and the role of technology in twenty-first century societies. Design is increasingly questioning established assumptions and beliefs and emphasizing the need for broader societal considerations in the design and deployment of new technologies.

In the previous chapter, we surveyed engineering practice based on his experience in the field, specifically systems engineering. Batteau notes the central role of *representation* of the systems he was building and analyzing, not only as marking on a physical or virtual page, but as the "rather formalized descriptions of the systems, their functionality, and their flows of information." This presents an opportune segue to the introduction of design, a related but now distinct professional practice. *Representation* is a central aspect of both engineering and design. Through my experience as a design educator and design anthropologist, I have seen designers increasingly turn

A. Batteau and C. Z. Miller, *Tools, Totems, and Totalities*,
https://doi.org/10.1007/978-981-97-8708-1_5

to systems thinking using various types of models, frameworks, and other forms of representation as vehicles for communication, sensemaking, and consensus-building with stakeholders, clients, and team members. Models of the design process (aka *design thinking*) start with understanding context and users. The Entity-Relationship-Attribute-Function (ERAF) model for example is commonly used in design to explore and map the system and subsystems in which design is intended to occur. It extends the traditional Entity-Relationship (ER) model by incorporating attributes and functions providing the designer with a more comprehensive view of a system's structure and behavior.

The purpose of this chapter is to sketch the contours of design from a perspective that is beyond technological hegemony in which technology is not perceived as the best and final answer to every need or problem. This perspective takes a critical and nuanced view of technological hegemony and the role of technology in twenty-first century societies, questioning established assumptions and beliefs, and emphasizing the need for broader societal considerations in the deployment of new technologies. A perspective that seeks to imagine a society beyond technological hegemony is not antithetical to technology, nor does it conceive of design as inextricably enmeshed within technology. Rather, it perceives design as a fundamental human activity and impulse to make and reorder the world in which technology is a complex assemblage of physical artifacts, social practices, and systems of meaning (Law in Bijker et al., 1987). This stands in contrast to modernist assumptions about technology's awe-inspiring power, usefulness, and mystique, as well as the expectation that technology will always and consistently improve human lives.

In this chapter, technology is understood to include hardware (devices), software (programs and platforms), and firmware (a form of microcode or a program embedded into hardware devices to help them operate effectively). Sociologist John Law, known for his work in the field of Science and Technology Studies (STS) and the concept of heterogeneous engineering,[1] defined technology as.

> ...a family of methods for associating and channeling other entities and forces, both human and non-human. It is a method, one method, for the conduct of *heterogeneous engineering*, for the construction of a relatively stable system of related bits and pieces with emergent properties in a hostile or indifferent environment. (Law, 2012)

Law's broad socio-technical perspective emphasizes the complex, networked, and contingent nature of social reality which is important to our understanding and aligns with the evolving meaning of technology over time.[2]

From the perspective of Science and Technology Studies (STS), tangible and intangible technological artifacts are entangled in complex actor networks, dynamic webs of human and nonhuman actants and flows of exchanges (Latour, 2005). John

[1] Heterogeneous engineering as defined by Law is "the construction of a relatively stable system of related bits and pieces" that are shaped and assimilated into a network. The "product can be seen as a network of juxtaposed components." (1987, p. 113).

[2] We have discussed in previous chapters how the meaning of technology has evolved over periods of human history, from Ancient and Classical periods to Medieval and Renaissance, Industrial Revolution, Modern, Post-modern, and into the digitized twenty-first century.

Law (Law, 1992), as well as Michel Callon (2012), Steve Woolgar (2012), Madeleine Akrich (1992), and others, has been instrumental in developing Actor Network Theory (ANT), a theoretical and methodological approach in which humans and nonhumans are actors in the social and natural worlds in constantly shifting networks of relationships and dynamic exchanges. Conceiving of technology *not as an isolated thing, object, or device,* but as a participating agent in a holistic network of human and nonhuman actors reflects the goal of establishing *relationality,* an approach that is increasing in design, and is critical to understanding the perspective we articulate here.[3]

The increasing complexity of design's domains, interest in the "agency and social life of materials" (Clarke, 2018), and the holistic and systemic turn is refocusing design from a solution-focused practice to an interventionist, speculative, thought-provoking practice that explores opportunities and potentialities and their consequences. Practice here aligns with Chia and Holt's conceptualization that rather than a conscious, defined process or set of methods, practice is "like water to a fish swimming in it." They paraphrase Heidegger's description of practice as "expressions of a shared know-how and generally acquired discrimination that resists any attempt to fix or limit it completely." (2009/2011, p. 129).

Design's transition from object-centered to human-centered and (now) beyond highlights the evolving role designers play in shaping a yet-to-be-fully-articulated—paradigm that decenters technology. Design is empowered by its confluence with other disciplines such as anthropology, specifically with the integration of ethnography in developing concepts for example product or service "in use" rather than the traditional industrial design focus on *the object* as an isolated entity. There are increasing signs of this *relational* perspective "that sees people and their tools as mutual participants in a system." (Chesluk & Youngblood, 2024). Holism, the basic tenet of ethnography, systems thinking and ecology theory, which study the complex dynamics of ecosystems, and complexity science are perspectives that are based on *relationality*, the view that nothing exists in isolation, and all things are in relationship with others.

As technological artifacts and their consequences seem to advance beyond control, the urgency of ethical questions is increasing. When does a "user" become the "used"? Rather than *convivial tools* (Illich, 1973; Sanders & Stappers, 2014) that allow people to customize and shape their use, digital technologies dictate who has access, how access is granted, when, and under what conditions digital technologies can be used.

This new design perspective challenges the technological hegemony and prevailing beliefs, mystique, and assumptions surrounding a modernist view of technology. Although it does not disavow technology, it recognizes that technology's contribution to human progress *is largely imagined* and curated. As noted in our Introduction, "the enshrinement of technology as the salvation of the human condition, a remedy for illness, mortality, hunger, and discomfort, required an attitude that

---

[3] An interesting change in the way products were represented in Industrial Design students' portfolios took place when humans and other actors were included with devices to portray the "thing in use" rather than a static rendering of the thing itself.

trusted efficiency, privileged the individual, and distrusted society – in short, a modern sensibility." (Introduction p. 1) Alternatively, a design perspective and approach that critically examines the role of moral imagination in shaping our perception and design of technology and anticipates its impact on society and the planet. In contemporary parlance, it seeks to consider "guardrails" and questions the belief in progress which is deeply rooted in modernist assumptions that prioritize efficiency and individualism. As we note in the Introduction, the critique of technology has a long history. More recently, the study of *values-based design* has raised the moral implications in the design of technology. In *Value Sensitive Design: Shaping technology with moral imagination,* Friedman and Hendry (2019) state:

> Technology is the result of human imagination - of human beings envisioning alternatives to the status quo and acting upon the environment with the materials at hand to change the conditions of human and non-human life. As a result of this human activity, all technologies to some degree reflect, and reciprocally affect, human values. It is because of this deep-seated relationship that ignoring values in the design process is not a responsible option. At the same time, actively engaging with values in the design process offers creative opportunities for technical innovation as well as for improving the human condition. (p. 1)

In the following sections, we draw on the work of designers, design thought leaders, and the speculative, moral, speculative and experimental dimensions of imagination to consider the roles that design might play beyond the current fixation on "high technology" that permeates modernist and post-modern societies. We address several fundamental questions: What is design? Who is a designer? What is the meaning and role of technology in relation to design?

The contours of a design perspective that looks beyond technological hegemony begin to take shape as we question where to locate the edges and how to set the guardrails on rapidly emerging technological platforms.[4] As this landscape continues to materialize, the implications are that designers will need to adapt to a new mindset and a sense of agency that require new knowledge, skills, and tools (Pendleton-Jullian & Seely Brown, 2018).

Design as Inquiry and Modes of Logic

Design is inextricably bound up within our ever-changing and often contested understanding of what it means to be human. Individual and societal level philosophical questions that ask, "Who are we?" and "Where do we fit in the universe?" As noted in Chap. 8, there are no definitive answers to these questions. Yet design reflects how particular societies and individuals construe their *particular* identities and cosmologies across space and time.

---

[4] "The 2024 Edelman Trust Barometer reveals a new paradox at the heart of society. Rapid innovation offers the promise of a new era of prosperity, but instead risks exacerbating trust issues, leading to further societal instability and political polarization." https://www.edelman.com/trust/2024/trust-barometer.

Depending on the context, design has been defined in various ways.[5] The following are not exclusive, but are common characteristics in many definitions of design:

- Design is purposeful and intentional.
- Design at its core is about identifying problems and seeking solutions.
- Design is interventionist and future oriented.
- Design is rarely linear. It is an iterative process of research, feedback, and refinement.
- As a process, design can be applied at a tactical level or strategically.
- Design is concerned with form (how something looks) and function (how it works).
- Design has evolved from object-centered to human-centered. With the emergence of sustainable design, the turn to systems thinking, and concerns about consequences, design is beginning to embrace a holistic perspective that extends beyond human-centeredness.
- Design is closely tied to creativity and innovation.
- Design is multidisciplinary incorporating various disciplines including social science, engineering, psychology, business, and others.

Pendelton-Jullian and Seely Brown would add that design is opportunistic and optimistic,[6] characteristics that have drawn criticism[7] from both within and outside the field (Suchman, 2011). "Design has always been a visionary pursuit and a visionary practice…The link between vision as a mental activity (imaging a future) and its accomplishment in the world (the building of the imagined future) is design." (Pendelton-Jullian & Seely Brown, 2018: 25) Design has learned to be responsive and adaptive to working in contexts that are increasingly complex, nuanced, and connected. Through training and *praxis,*[8] designers develop the ability to engage in environmental scanning, consciously and unconsciously updating their perceptions and conceptions of the worlds in which they live and work. Individually and collectively, they develop *system wisdom* (Chia & Holt, 2009/2011) through interactions with people, places, and things that make up the networks of their lived experiences.[9]

---

[5] Thanks to Jordan Pollack for his insight about the connection between "envisioned outcome with world-adapted accomplishment. Searle (1984) famously frames this in intentional terms as the "mind-to-world direction of fit" structuring desire and action (in contrast to the world-to-mind fit constituting perception and belief).".

[6] The authors note that "the optimism associated with design is a **skeptical optimism**." (p. 26).

[7] Lucy Suchman has been a vocal critic of design hubris, what might be called design arrogance, specifically calling out Bruce Mau's (2004) uncritical view of the role of design over time.

[8] "By *praxis* we mean the 'conduct of one's life and affairs primarily as a citizen of the polis; it is activity which may leave no separately identifiable outcome behind it and whose end, therefore, is realized in the very doing of the activity itself.'" (Dunne, 1993, as cited by Pendleton-Jullian & Seely Brown, 2018).

[9] "What is known of a system and our place within it, therefore, is immanent not to some part of a system but to systems within systems, the edges of which can never be closed. For Bateson, it was our human tendency to forestall this immanence by constantly looking for end points, for definitive assessments, for neat theories, for well-conceived and easily explicable earning opportunities that, somewhat ironically, cre ated a lack of awareness." (Chia & Holt, 2009/2011: 89) The authors are referring to anthropologist Gregory Bateson.

Beyond conceiving of design within the context of commercialism, business, and industry, the intention here is to consider design in broad terms: its evolution over time, and how it might impact and be impacted by the material and non-material contours of a post-technological perspective that challenges prevailing assumptions and beliefs surrounding technology which, in mainstream culture, are varied, often contradictory, and represent diverse perspectives.

Design is a human impulse and activity that is not limited to a single discipline, academic field, domain, or arena of practice. Design cannot be limited to a single occupation: *designer*. Professional certification is available from academic or other institutions that provide training in a growing number of sub-fields and associated skill sets. Despite the professionalization of design in the form of certification and academic degrees, attempts to limit those who have a claim to the title *designer* is problematic. While recognizing education and training that result in the designation of *professional designer*, the intent here is to broaden the aperture. Rather than restricting admittance, we refer to previously noted common characteristics of design expressed through various definitions and sets of activities to establish our criteria. We use these criteria to sketch the broader, more inclusive parameters of *design practice* and what it means to identify as a *designer*.

Designers typically work within multiple constraints: specific client requirements, available materials, contextual situations such as physical environment, anticipated users[10] (Hale, 2018), and the enabling and restraining forces of culture. Culture is a continuously shifting rubric of values, beliefs, symbols, and materiality expressed through multiple dimensions that constitutes the contextual parameters within which design operates. At the same time, designers are dependent on the capacity to explore multiple realms of imagination which they do through *abductive reasoning*, a mode of logic and way of thinking to which designers are particularly attuned. Abductive reasoning seeks to generate *plausible premises*[11] that are not definitively verified, do not eliminate uncertainty, and allow for iteration. Interventions and solutions are presented as "best available" or "most likely".[12]

[10] We refer here to Tamara Hale's article "People Are Not Users" in the special issue of *Journal of Business Anthropology* on design anthropology. Hale argues that referring to individuals as "users" ignores their diversity, deletes the social aspects of their lives and vastly underestimates the complexity of those who "use" products and services.

[11] Thanks to Jordan Pollack for suggesting the change from "plausible conclusions" to "plausible premises." In fact, abductive reasoning is about possibilities or "what Peirce termed simply the "method of hypothesis." Premises, what are referred to in design as principles or "design criteria" are used to develop concepts that are then tested to validate a potential 'solution'.

[12] Abductive logic was first articulated by Charles Sanders Peirce in the late nineteenth century. It is conceived as an alternative to more established forms of logic: deductive and inductive reasoning. https://en.wikipedia.org/wiki/Abductive_reasoning. Accessed July 10, 2023.

Abductive Reasoning in Design

Abductive inference,[13] a logistical mode formulated by Charles Sanders Pierce (1839–1914) in the late nineteenth century, is a key characteristic of design and the design process, widely known as *design thinking*. Unlike deductive reasoning, which infers if the premises are true, then the conclusion must be true[14] or inductive reasoning that infers a general principle based on a body of knowledge that holds that the general principle is *likely* to be true[15] abductive reasoning seeks the "most likely explanation" or "best likely" solution or decision thereby acknowledging a higher degree of uncertainty, the proverbial "leap of faith."

What is commonly known as "design thinking" draws on the abductive reasoning process that, in essence, involves five modes of inquiry: observation, pattern recognition, hypothesis formulation, prototype development, and user testing. There are several well-known models of the abductive reasoning process applied in design, e.g., the Stanford d.school,[16] Vijay Kumar's design innovation process model (Kumar, 2013), and the Double Diamond[17] or "Bowtie" model.

In the problem-solving mode of design, constraints on designers originate from several sources such as a client brief or problem to be solved, the project timeframe, resources, and budget. Tight constraints can result in incremental change rather than innovative interventions. The tensions implicit in the iterative design process are rarely if ever completely resolved. Stages of the design process that support abductive reasoning such as ethnographic observation and contextual research might be shortened or entirely passed over. The delivered product, the result of *iterative refinement*, is a beta prototype, theoretically open to revision with additional information and resources, including time.

An example of a design project that resulted from the abductive reasoning process is the Dyson vacuum cleaner developed by British inventor and engineer, James Dyson. Dyson set out to design a more effective and user-friendly vacuum cleaner that could maintain constant suction and eliminate the need for bags. Although the terminology differs, Dyson's adductive process basically followed the Stanford design thinking model (Fig. 5.1): *observe, identify patterns, formulate hypotheses, develop design criteria, prototype development, and user testing*. The observation phase began with Dyson's personal irritation with conventional vacuums that tended to clog as the bag became full. Having identified the limitations of existing vacuum cleaners, Dyson recognized a pattern that he traced back to the reliance on bags to capture

---

[13] Inferences are steps in reasoning that lead from premises to conclusions.

[14] An example of deductive inference from the premises "all men are mortal" and "Socrates is a man" therefore "Socrates is mortal.".

[15] An example of inductive reasoning: if all the swans observed so far are white, then it is probable that all swans are white. The study of a particular population is generalized to a larger population.

[16] Sanford d.school design thinking model https://empathizeit.com/design-thinking-models-stanford-d-school/.

[17] British Design Council (2003) Double Diamond model https://www.designcouncil.org.uk/our-resources/the-double-diamond/.

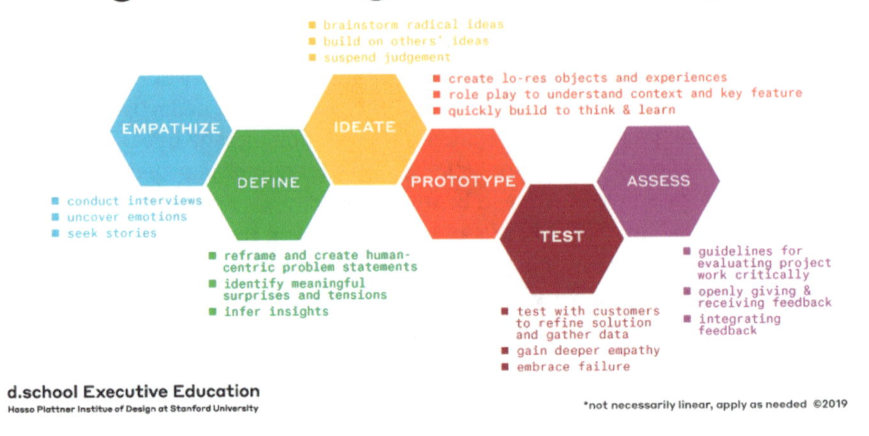

Fig. 5.1 5 steps design thinking model proposed by the Hasso-Plattner Institute of Design at Stanford d.school

dust and debris. He then formulated an initial hypothesis around the idea of using cyclonic technology, inspired by the principle of centrifugal separation, to ensure continual suction and reduced clogging. After developing a set of design criteria, Dyson set out to design and test prototypes. Finally, he put prototypes into the field for several rounds to test them under various conditions in real-life situations. He collected feedback from tests conducted with potential users. The feedback was used through several iterations of refinement to produce the final design of the product.

Abductive reasoning provides a work-around for more rigid forms of logic and opens the possibility for speculation in the cognitive process spectrum proposed by Pendleton-Jullian and Seely Brown (2018). Speculation is an exploration of things that *might be* (Liedtka & Ogilvie, 2011*)* but that are grounded in what is already sensed and seen as emerging. Speculative design is a methodological approach that explores potential futures by creating design concepts, artifacts, and scenarios that are not only problem-solving exercises, but are meant to challenge and provoke our assumptions about the way things are—"what is"—and to stimulate debate about the kind of future we want to live in. Speculative design demonstrates the major role the convergence of design and anthropology plays in shaping human experience and society not only in present circumstances, but also in possible futures "as an extension of the ethnographic gaze." (Halse, 2013) Rather than operating in service to corporate interests, speculative design explores possibilities beyond those that address existing human needs. The design anthropological practice of speculation does not intend to predict, but instead to create a space for exploring possible futures. By creating hypothetical future scenarios based on emerging trends in technology, science, politics, and culture, speculative design creates a space for examining different possible future states and their implications and potential consequences.

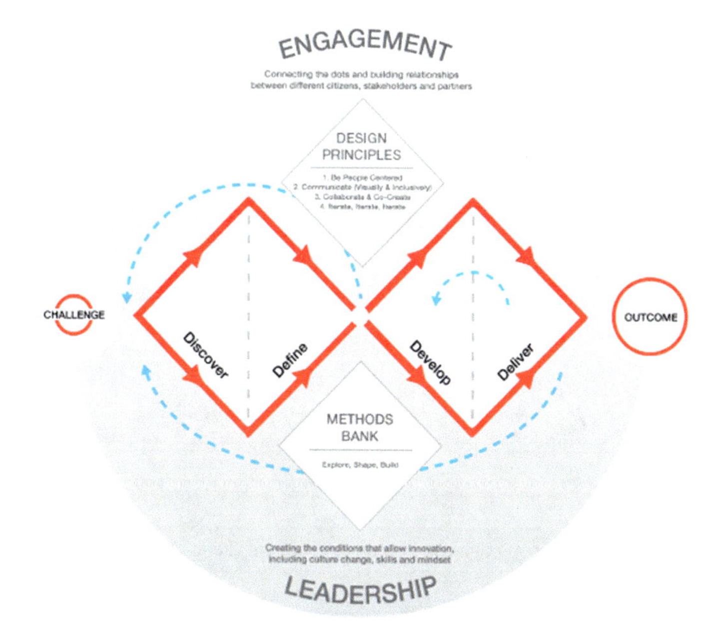

Fig. 5.2 Design Council double-diamond model (licensed under a CC BY 4.0 license)

Figure 5.3 illustrates the trajectory from perception to modes of reasoning including deductive, inductive, and abductive logic. Accessing the realms of imagination beyond modes of reasoning requires freedom from prescribed constraints. Beyond abductive logic, designers explore potentiality and possibility through the cognitive processes of *imagination* that include speculation, experimentation, and free play (Fig. 5.2).

In *Design Unbound*, Pendleton-Jullian and Seely Brown (2018) identify a set of initial principles that cover the range of imagination (pp. 430–431):

1. The imagination serves diverse cognitive processes as an entire spectrum of activity.
2. The imagination both resolves and widens the gap between the unfamiliar—the new/novel/strange—and the familiar. This gap increases along the "role of imagination in cognitive processes" spectrum from left to right.
3. Pragmatic Imagination proactively imagines the action in light of meaningful, purposeful possibilities, and sees the opportunity in everything.
4. The Pragmatic Imagination sees thought and action as indivisible and reciprocal. Therefore, it is part of all cognitive activity that serves thought and action for anticipating and thought and action for follow-through. The generative/poïetic/sometimes-disruptive side of the spectrum is especially critical in a world that requires radically new visions and actions.
5. The entire spectrum of imagination must be instrumentalized to turn ideas into action.

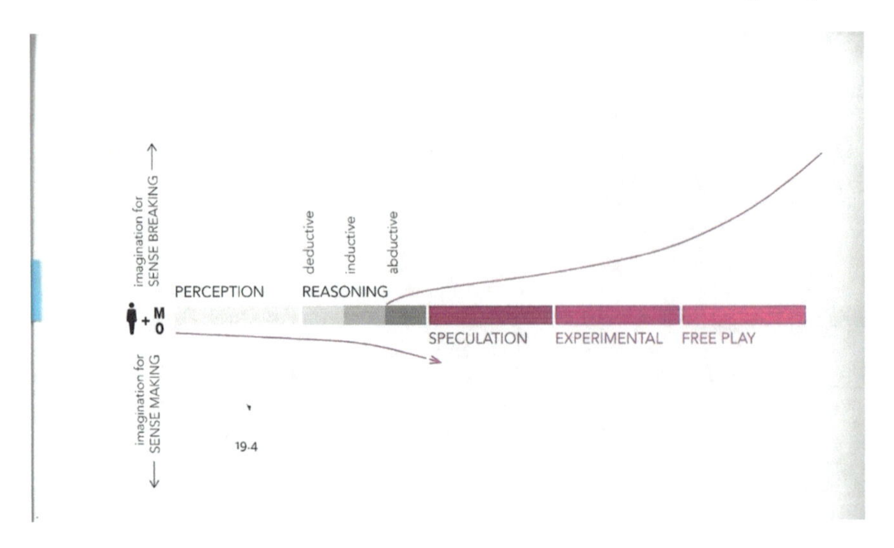

Fig. 5.3 Role of imagination in cognitive processes spectrum (Pendleton-Jullian & Seely Brown, 2018, p. 410)

6. Because the imagination is not under conscious control, we need to find and design ways to set it in motion and scaffold it throughout meaningful activity.

Accessing imagination by means of abductive reasoning allows designers—and creatives in general—to experiment and explore potential future trajectories. Design is often associated with the "generative/poïetic/sometimes-disruptive side of the spectrum". However, it is widely recognized within the field that as the entire spectrum of imagination is activated (e.g., principle 4), an awareness has become more explicit over time.

In summary, design is an intentionally interventionist, transformative, and transdisciplinary practice that has evolved beyond a "late stage add-on", an object-oriented and craft-based set of skills (Brown, 2009, p. 1). Contemporary design is positioned squarely within the domain of innovation as a creative force and an agent of both disruption and renewal. Design today is concerned with facilitating innovation and the intentional creation of "The New" (Erwin, 2014). Although the primary role of design has been to facilitate the innovation process in service to corporate goals, design has demonstrated a capacity for creative problem-solving in an increasing range of domains including but not limited to health care, education, civic engagement and governance, workflow and process design, experience design (i.e., entertainment and user experience) service design, and design management. Courses focused on ethics and sustainability in design programs suggest a growing sense of moral responsibility (Lima, 2023; de Mozota & Amland, 2020; Papanek, 1973) that aligns with key aspects of a post-technological perspective, specifically, the shift to holistic, system thinking, consideration for consequences, and a more nuanced and critical approach that decenters technology.

In the section that follows we introduce designers whose work illustrates aspects of what a post-technological perspective in the practice of design is likely to entail. In several cases, we include principles developed by thought leaders such as Rittel and Webber (1973) and Buchanan (1992) that sketch the contours of an evolving design paradigm that calls for *convivial technology* that supports the ethical values and design criteria that can be summed up on five dimensions: relatedness, adaptability, accessibility, bio-interaction, and appropriateness (Vetter, 2017).

References

Akrich, M. (1992). The de-scription of technical objects. In Bijker, W, Law, & J (Eds.), *Shaping technology/building society. Studies in sociotechnical change* (pp. 205–224). MIT Press. https://shs.hal.science/halshs-00081744
Bijker, W. E., Hughes, T. P., & Pinch, T. (Eds.). (1987). *The social construction of technological systems: New directions in the sociology and history of technology*. MIT Press.
Biton, A., Shoval, S., & Lerman, Y. (2022). The use of cobots for disabled and older adults. In *14th IFAC Workshop on Intelligent Manufacturing Systems IMS 2022*
Boylston, S. (2019). *Designing with society: A capabilities approach to design, systems thinking, and social innovation*. Taylor & Francis.
Boylston, S. (2023). *Exploring nature inclusive design principles* LinkedIn.
Brown, T. (2009). *Change by design*. Harper Collins Publishers.
Buchanan, R. (1992). Wicked problems in design thinking. *Design Issues, 8*(2), 5–21. https://doi.org/10.2307/1511637
Callon, M. (2012). Society in the making: The study of technology as a tool for sociological analysis. In W. E. Bijker, T. P. Hughes, & T. J. Pinch (Eds.), *The social construction of technological systems: New directions in the sociology and history of technology* (pp. 83–103). MIT Press.
Chesluck, B., & Youngblood, M. (2024). Systems, complexity, and human experience in design anthropology: From user-centeredness to user ecosystem thinking. In C. Z. Miller & J. M. Spears (Eds.). *The Routledge Companion of Practicing Anthropology and Design*. Routledge Taylor & Francis.
Chia, R. C. H., & Holt, R. (2009/2011). *Strategy Without Design: The Silent Efficacy of Indirect Action*. Cambridge University Press.
Clarke, A. J. (Ed.). (2018). *Design anthropology: Object cultures in transition*. Bloomsbury Academic.
De Mozota, B. B., & Valade-Amland, S. (2020). *Design: A business case*. Business Expert Press.
Edelman. (2024). 2024 Edelman trust barometer. Retrieved from https://www.edelman.com/trust/2024/trust-barometer
Friedman, B., & Hendry, D. G. (2019). *Value sensitive design: Shaping technology with moral imagination*. MIT Press.
Hale, T. (2018). People are not users. *Journal of Business Anthropology, 7*(2), 163–183.
Halse, J. (2013). Ethnographies of the possible. In T. O. Wendy Gunn, and Rachel Charlotte Smith (Ed.), *Design anthropology: Theory and practice*. Bloomsbury.
Illich, I. (1973). *Tools for conviviality* (Vol. 47). Harper & Row.
Kallis, G. (2018). *Degrowth*. Agenda Publishing.
Kim, H. H. (2023). A.I. is a Fiction: Stop freaking out when chatbots say they're in love or make disturbing threats. Just treat them like Pinocchio. *Wired*, (31.09).
Kumar, V. (2013). *101 design methods: A structure approach for driving innovation in your organization*. John Wiley & Sons.

Latour, B. (2005). *Reassembling the social: An introduction to actor-network-theory.* Oxford University Press.

Latour, B. (2006). Mixing humans and nonhumans together: The sociology of a door-closer. In W. B. William, A. H. Jonathan, & J. S. Sadar (Eds.), *Rethinking technology: A reader in architectural theory* (p. 16). Routledge.

Law, J. (1992). Notes on the theory of the actor-network: Ordering strategy and heterogeneity. *Systems Practice, 5*, 379–393.

Law, J. (1987). Technology and heterogeneous engineering: The case of portuguese expansion. In W. E. Bijker, T. P. Hughes, & T. Pinch (Eds.), *The social construction of facts and artifacts: New directions in the sociology and history of technology.* MIT Press (1987) (Reprinted from 1999)

Law, J. (2012). Technology and heterogeneous engineering: The case of the portuguese expansion. In J. Law, T. Hughes, T. Pinch, & W. Bijker (Eds.), *The social construction of technological systems* (Anniversary ed.). MIT Press.

Liedtka, J., & Ogilvie, T. (2011). *Designing for growth: A design thinking toolkit for managers.* Columbia University Press.

Lima, M. (2023). *The new designer: Rejecting myths.* MIT Press.

Lima, M. (2011). *Visual complexity: Mapping patterns of information.* Princeton Architectural Press.

Lima, M. (2014). *The book of trees: Visualizing branches of knowledge.* Princeton Architectural Press

Mau, B. (2004). *Massive change* Phaidon Press.

Meadows, D. H. (1972). *The limits to growth: A report for the club of Rome's project on the predicament of mankind.*

Meadows, D. (1999). *Leverage points: Places to intervene in a system.*

Meadows, D. (2008). *Thinking in systems: A primer*. Chelsea Green Publishing.

Mori, M. (1970). Bukimi no tani [The uncanny valley]. *Energy, 7*, 33.

Mori, M., MacDorman, K. F., & N. Kageki. (2012). The Uncanny Valley [From the Field]. *IEEE Robotics & Automation Magazine, 19*(2), 98–100. https://doi.org/10.1109/MRA.2012.2192811

Papanek, V. (1973). *Design for the real world: Human ecology and social change.* Bantam Books.

Pendleton-Jullian, A. M., & Brown, J. S. (2018). *Design unbound: Designing for emergence in a white water world* (Vol. 1). MIT Press.

Reznich, C. (2017, July 17). 1973 horst Rittel.

Rittel, H., & Webber, M. M. (1973). Dilemmas in a general theory of planning. *Policy Sciences, 4*(2), 155–169.

Sanders, E. B.-N., & Stappers, P. J. (2014). *Convivial Toolbox: Generative research for the front end of design.* BIS Publishers.

Searle, J. R. (1984). Intentionality and its place in nature. *Dialectica, 38*(2–3), 87–99.

Suchman, L. (2011). Anthropological relocations and the limits of design. *Annual Review of Anthropology, 40*, 1–18.

Vetter, A. (2017). The matrix of convivial technology—Assessing technologies for degrowth. *The Journal of Cleaner Production, 197*.

Wang, T. (2021, June 2021). The most popular strategy in design thinking is BS *Fast Company.*

Woolgar, S. (2012). Configuring the user: The case of usability trials. In W. E. Bijker, T. P. Hughes, & T. J. Pinch (Eds.), *The social construction of technological systems: New directions in the sociology and history of technology* (pp. 57–99). MIT Press.

Zachary, G. P. (1977). *Endless Frontier: Vannevar Bush.* Free Press.

Chapter 6
Design for Convivial Technology

Abstract In this chapter, we trace the emergence of modern design starting with the German Bauhaus movement and then to influential design thought leaders. Some of these individuals were trained as professional designers, while others come from various disciplines that suggest the contours and characteristics of design as it leaned toward becoming an interdisciplinary field, moving away from its traditional object orientation, and opened to consider the realms of experimentation and speculative imagination. We see how designers began to exhibit agency especially when not in the service of industry and commercial interests. Finally, we explore the concept of conviviality and the potential for design to lead in envisioning and enacting an approach toward technology that moves away from the tight coupling (which we discuss in Chap. 12) and control that contributes to technological hegemony.

In the previous chapter, we outlined the links between design and its role in technological development. Beyond its perception as craft, the evolution of design as a field was influenced by the German Bauhaus movement (1919–1933) which greatly contributed to the professionalization of design and the formalization of design education. Even after its closure by the Nazi regime the teachings and principles of the Bauhaus continued to influence artists, designers, and architects across the world (Fig. 6.1).

Many Bauhaus members emigrated and spread its ideas internationally, especially to the United States,[1] where they had a hand in shaping the development of American modernism and many modern design principles. Their iconic designs captured the

[1] Founded as an art school by architect Walter Gropius in Weimar, Germany, many members of the Bauhaus relocated to the U.S. after the closure of the school in Berlin in 1933. Gropius was installed at Harvard, László Moholy-Nagy founded the New Bauhaus in 1937 in Chicago at what became the Institute of Design at the Illinois Institute of Technology and was joined at ID by Mies van der Rohe. Josef Albers joined the faculty at Black Mountain College and was eventually hired at Yale.

Convivial, or Conviviality: **con·viv·i·al (/kənˈvivēəl/)** the ability of individuals to interact creatively and autonomously with others and their environment to satisfy their own needs.

A. Batteau and C. Z. Miller, *Tools, Totems, and Totalities*,
https://doi.org/10.1007/978-981-97-8708-1_6

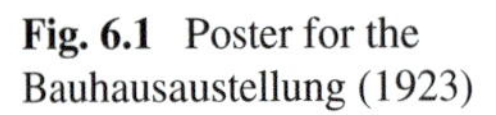

Fig. 6.1 Poster for the Bauhausaustellung (1923)

soul of twentieth century technology, a glorification of form, function, and power. Coming out of the Industrial Revolution, this new conceptualization of technology was broader and more complex, encompassing not just machinery and industrial processes but also electronics and information systems. Bold designs in angles and sensuous curves exuded speed, efficiency, and signaled the arrival of technological dominance.

In this chapter, we introduce the concept of *conviviality*, a common characteristic of an alternative approach to technology that promotes design and designing as a social activity rather than an industrial or commercial one. Conviviality was explored by Ivan Illich in *Tools for Conviviality* (1973). Liz Sanders and Pieter Jan Stappers applied the concept to generative design research which they describe as, "an approach to bring the people we serve through design directly into the design process in order to ensure that we can meet their needs and dreams for the future." (2014, p. 6).

Beginning with Victor Papanek (1923–1998), each of the design thought leaders featured in this chapter posed a challenge to the hegemony of modernist technology that was represented in much of twentieth and twenty-first century design. Each contributes insights and critiques, and advocates for different perspectives and methodologies that decenter technology as the dominant actor in the modernist narrative of design and innovation. The contours of a *convivial design* approach to technology begin to take shape in the sample of designers and design thought leaders whose work embodies the principles and values of an alternative paradigm and narrative. The common themes reveal the framework for a path forward that supports and sustains the diversity and inclusion of life on this planet: acknowledging the totality—the whole of life and the relationality between all that exists, recognizing the limits of the dominant for profit economic model (i.e., late-stage capitalism), decentering technology and rebalancing its role as a convivial tool. And most importantly, a sense of optimism that drives the impulse to design and create.

The sequence of design thought leaders introduced in this section is not intended to be exclusive. Some are professional designers, others whose training and practice is centered in other disciplines. The intention is to show how the connections and overlaps between these individuals begin to emerge as a loosely structured network. We could include many other thought leaders and designers who have had a direct impact on the evolution of design, who have either influenced or been influenced by those selected to represent an emerging paradigm. Rather than a comprehensive overview, these brief sketches provide a limited sample of work that begins to map the contours of what a convivial, integrative, and holistic design might entail.

A common characteristic of this sample of design thought leaders from the twentieth to twenty-first century is the sense of *agency,* the capacity of individuals to act independently and make their own free choices. As a construct, agency has been studied extensively by scholars from many disciplinary perspectives. A short list includes philosophy, Kant (1788) and Sartre (1943); in sociology, Giddens (1984) and Bourdieu (1977); in economics, Becker (1976); and in psychology Bandura (1989). More recent perspectives include feminist scholar Judith Butler (1990), cognitive and neuroscientist Benjamin Libet (1999), and anthropologist Sherry Ortner (2006).

Related to the idea of agency, these thought leaders share the perception of *conviviality*, an alternative approach to designing that promotes flexibility and the intentions, creativity, and purposes of people (aka "users"). This is design that exists mainly outside design's service to corporate interests. Conviviality as a *design principle* is a thread running through the ideas of everyone featured in this section. This is design "after hours", apart from the day job of a designer employed in industry or after teaching and office hours in university and design schools.

Victor Papanek

No discussion of contemporary design thought would be complete without considering the work of designer and educator, Victor Papanek (1923–1998). Papanek is widely known for his position on socially responsible design. He was unabashedly critical of the profession of industrial design. In the Preface to a revision of *Design for the Real World* Papanek stated.

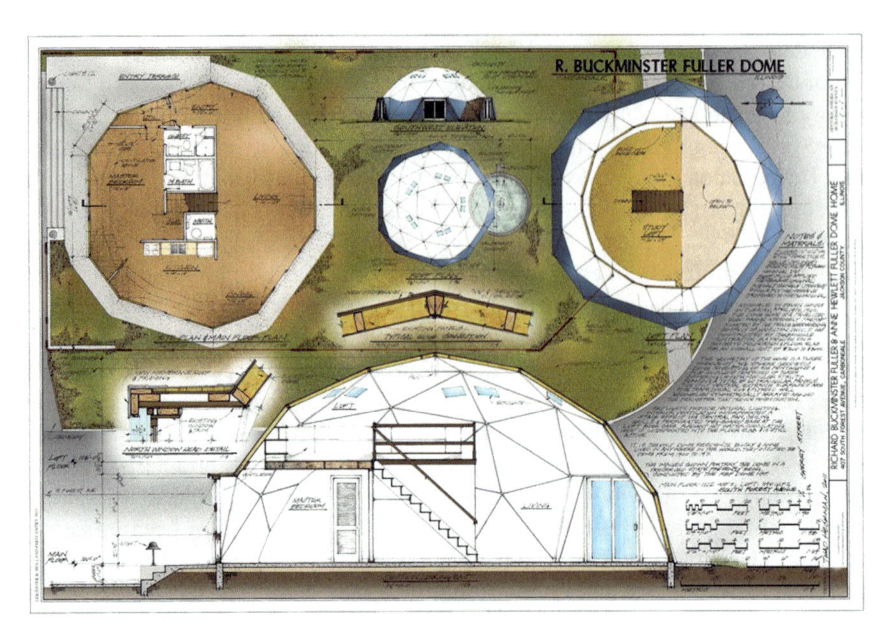

Fig. 6.2 Richard Buckminster Fuller and Anne Hewlett Fuller Dome Home, 407 South Forest Avenue, Carbondale, Jackson County, IL Drawings from Survey HABS IL-1234

> There are professions more harmful than industrial design, but only a very few of them. And possibly only one profession is phonier. Advertising design, in persuading people to buy things they don't need with money they don't have, in order to impress others who don't care, is probably the phoniest field in existence today. (1973, p. 19)

Commenting on the technology of the time Papanek's longtime friend and collaborator, inventor and humanitarian, R. Buckminster Fuller (1895–1972), noted in the Introduction to *Designs for the Real World*:

> At M.I.T. there are buildings full of rooms, and rooms full of yesterday's top priority machinery that is now utterly obsolete. They have a vast graveyard of technology. The students don't want to take classes in mechanical engineering any more because they have heard that what they learn is going to be obsolete before they graduate. These evolutionary events cover all phases of technology and the physical sciences. Victor Papanek's book conducts a mass funeral service for a whole segment of now obsolete professionals. (1973, p. 10)

Although both Papanek and Fuller are known within design and beyond, their warnings have gone mostly unheeded. The iconic geodesic dome[2] perhaps Fuller's best-known work, was recognized as an impressive feat of engineering. However, it never materialized as a commercial success on a large scale (Fig. 6.2).

---

[2] Buckminster "Bucky" and Anne Hewlett Fuller's Dome home in Carbondale, Illinois is listed on the U.S. National Register of Historic Places.

Papanek and Fuller continue to be cited as sources of inspiration for those who make the case for an alternative vision of design that is not bound to purely commercial interests.

Horst Rittel and Melvin Webber: A New Class of "Wicked" Problems

Horst Rittel (1930–1990) is best known for his concept of "wicked problems" (Rittel & Webber, 1973) which was developed in collaboration with Melvin M. Webber (1920–2006), his colleague at the University of California at Berkeley. However, Rittel's influence on design theory and practice was much wider. Trained in mathematics and physics, as a Professor of Design Methodology at the Ulm School of Design in Germany from 1958 to 1963 (Hochschule für Gestaltung—HfG Ulm), Rittel taught science and engineering principles to designers. His classes on information and communication theory and the philosophy of science connected the latest in scientific discovery with the design process.[3]

It is important in understanding Rittel's role in the evolution of design thinking and practice to segue to the period following WWII. Recall the Stanford Dschool "Design Thinking Process Diagram" (Fig. 4.1) compared to Claude Shannon's model in Fig. 6.3, an example of the types of models that were common in the post-WWII period. Shannon was a student of Vannevar Bush (Bush, 1925/2021), the American engineer, inventor and science administrator who served as director of the U.S. Office of Scientific Research and Development (OSRD) during WWII and was influential in the Manhattan Project. Perhaps it is not surprising that Shannon's model is reminiscent of the engineering-inspired linear R&D (Research and Development) model developed by Bush[4]. Although highly influential in science, engineering, and computer science, Bush was not an advocate of the social scientists of his day. He was adamant that the National Science Foundation (NSF), which he supported, strip out programs related to the social sciences. G. Pascal Zachary notes Bush's strong support for the hard sciences including such as engineering and physics shaped what was judged as "science" and what was not. He was steadfast in his opposition to funding humanities and social science research and was reported to comment "I have a great reservation about these studies where somebody goes out and interviews a bunch of people and reads a lot of stuff and writes a book and puts it on a shelf and nobody ever reads it." (Zachary, 1977) Bush's lack of vision in understanding the value of social science research has had a chilling effect that persists today.

These often-overlooked aspects of the historical record are important in attempting to map out what a post-technological design might entail and where its roots, mostly undiscovered, might lie. In retrospect, Rittel and others like him played a critical

---

[3] Christopher Reznich https://medium.com/@creznich

[4] For a detailed discussion on Bush's the role refer to Technology and the Common Good: The Unity and Division of a Democratic Society (Batteau, 2022).

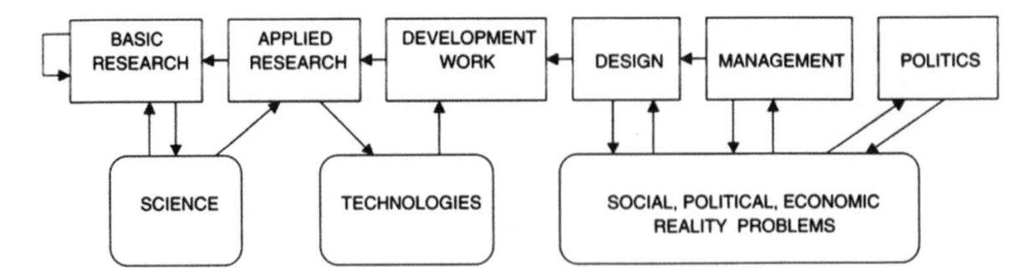

Fig. 6.3 "From the technical details of Claude Shannon's information theory to the newest research on cybernetics, Rittel connected the newest in scientific discovery with the design process." (Reznich, 2017)

role in maintaining an alternative to Vannevar Bush's point of view. *Dilemmas in a General Theory of Planning* open by describing the public's discontent with "professionals" and policy makers that fail to solve the problems that confront society. Rittel and Webber made a critical observation in explaining that failure is inevitable *due to the nature of the problems themselves*:

> The search for scientific bases for confronting problems of social policy is bound to fail, because of the nature of these problems. They are "wicked" problems, whereas science has developed to deal with "tame" problems. Policy problems cannot be definitively described. Moreover, in a pluralistic society there is nothing like the undisputable public good; there is no objective definition of equity; policies that respond to social problems cannot be meaningfully correct or false; and it makes no sense to talk about 'optimal solutions' to social problems unless severe qualifications are imposed first. Even worse, there are no "solutions" in the sense of definitive and objective answers. (1973, p. 155)

Rittel and Webber are important to this discussion because they identified a new class of problems that challenged the optimistic belief that the shortcomings of modern society could be solved through modern science by "the professions" (experts), each a specific subset of engineering, through which scientific knowledge is applied. The belief that planning for betterment is possible would not be abandoned. No matter how often implementation of the plans, which often included significant investments in technology, resulted not only in exacerbating the existing problem but also led to unanticipated consequences that then required another plan to fix the new problem. Development efforts by the World Bank in Africa and Indonesia to increase crop yields provide examples of projects that failed to consider the social context into which programs were being introduced. The plan in Kenya was designed to offer farmers the opportunity to increase agricultural production and their thereby economic position by growing and selling cash crops. The program provided farmers in poorer rural areas of Kenya a loan package consisting of money and seed-fertilizer combinations for corn, tobacco, and coffee. Farmers who accepted the loans were expected to sell the crops to the local farming cooperative to repay the loan over time. In the process, they would theoretically create wealth for their family, the region, and the country. However, based on the rate of repayment, the project was considered a complete failure. The project failed for several reasons that reveal a lack of understanding of the environmental and social context, and of the farmers themselves. For example, it was expected that farmers would embrace the opportunity to be agricultural entrepreneurs by growing and selling cash crops. This was not the case. In reality, their ambitions lie elsewhere.

Despite the vicious cycle of ill-advised development projects, Rittel and Webber note that failures of modern policy science have provided them with learning opportunities.

> We have been learning to ask whether what we are doing is the *right* thing to do. That is to say, we have been learning to ask questions about the *outputs* of actions and to pose problem statements in valuative frameworks. We have been learning to see social processes as the links tying open systems into large and interconnected networks of systems, such that outputs from one become inputs to others. In that structural framework it has become less apparent where problem centers lie, and less apparent *where* and *how* we should intervene even if we do happen to know what aims we seek. We are now sensitized to the waves of repercussions generated by a problem-solving action directed to any one node in the network, and we are no longer surprised to find it inducing problems at greater severity at some other node. And so we are forced to expand the boundaries of the systems we deal with, trying to internalize those externalities. (1973: 159)

These "externalities"—the complex interactions between components of context that were overlooked or ignored by policy professionals—are what Papanek and Fuller called out in their warning to designers: "As socially and morally responsible designers, we must address ourselves to the needs of a world with its back to the wall while the hands on the clock point perpetually to one minute before twelve." (1973, p. 19) Fifty years later, the warning is more urgent than ever as the clock moves relentlessly toward twelve.

In Dilemmas in a General Theory of Planning, Rittel and Webber (1973: 161–167) outlined 10 characteristics of the new class of problems that they characterize as "wicked problems" that differ from simple problems in that they are multifaceted, interconnected, and complex. They defy solutions and are often exacerbated, often spawning new problems when "solutions" are implemented. Examples of wicked problems such as climate change, poverty, and social injustice have the following characteristics in common:

1. There is no definitive formulation of a wicked problem.
2. Wicked problems have no stopping rule.
3. Solutions to wicked problems are not true-or-false but good-or-bad.
4. There is no immediate or ultimate test of a solution to a wicked problem.
5. Every solution to a wicked problem is a "one-shot operation"; because there is no opportunity to learn by trial-and-error, every attempt counts significantly.
6. Wicked problems do not have an enumerable (or an exhaustively describable) set of potential solutions, nor is there a well-described set of permissible operations that may be incorporated into the plan.
7. Every wicked problem is essentially unique.
8. Every wicked problem can be considered to be a symptom of another problem.
9. The existence of a discrepancy representing a wicked problem can be explained in numerous ways.
10. The planner has no right to be wrong.

Rittel and Webber argue that problems generated by large, complex, and interconnected networks of systems, systems within systems, must recognize a new *wicked* class of problems for which twentieth century policy science was not equipped. There are no silver bullets, technological or otherwise. The implications require that designers need to expand their mindset and practice beyond object-centered traditional design to a systems approach. What an expanded design mindset and practice would entail has taken decades to envision.

Richard Buchanan

Richard Buchanan is a Professor of Design and Innovation in the Weatherhead School of Management at Case Western Reserve University. He is a renowned scholar in the field of design, particularly for his work on design theory and design thinking. He is known as a pioneer in expanding the application of design beyond traditional practice, arguing that design thinking principles can be applied to complex problems across a wide range of areas.

One of Buchanan's most influential ideas is the "Four Orders of Design" which provides a framework to understand the broad scope of design practice:

1. Symbolic and Visual Communications: This includes traditional graphic design tasks like creating logos, symbols, book and magazine layouts, and signage.
2. Material Objects: This covers industrial and product design.
3. Activities and Organized Services: This concerns the design of interactions, processes, and services. Examples might include designing the process flow in a hospital, or the user experience of a digital service.
4. Complex Systems or Environments for Living, Working, Playing, and Learning: This is about the design of systems or environments, such as an eco-village or an educational system.

Buchanan's emphasis on the "Four Orders of Design" has been influential in expanding understanding and teaching of design as a discipline. Another of his contributions was to expand on Rittel and Webber's concept of "wicked problems." In the article "Wicked Problems in Design" (1992), Buchanan argues that based on the trends in design thinking over the twentieth century design "now should be recognized as a new liberal art of technological culture." (1992: 5).

In addition to his theoretical contributions, Buchanan has played a significant role in design education. He was head of the school of design at Carnegie Mellon University for several years and helped shape their curriculum around his expansive view of design. His work has significantly influenced the field of design and how design thinking is applied to solve complex problems in various domains. It is not only creating aesthetically pleasing and functional products, but also about improving experiences, services, and systems in a human-centered way.

Klaus Krippendorff: *The Semantic Turn*

Klaus Krippendorff (1932–2022) was a Professor of Communication at the University of Pennsylvania, a cybernetician, designer, and researcher. He received his education at the Ulm School of Design and earned a doctorate at the University of Illinois Champaign-Urbana. Krippendorff is known for his contributions to the emergence of human-centered design (HCD) and to advancing the science of design. Figure 6.4 depicts what he refers to as "distinctions within design practices." (Krippendorff, 2006, p. 32) Krippendorff distinguishes between three concepts of design: "design as the realization of everyday life", technology-centered design, and professional or "human-centered design" that is "derived from stakeholders' lives and made available in community." He identifies two types of professional design:

> Among professional designers, there are those concerned with strictly technological artifacts, with artifacts that may well be designed without consideration for their users' conceptions, by engineers for example and those concerned with human interactions generally and human interfaces with technological artifacts in particular. They are called here technology-centered and human centered designers, respectively. Technology-centered design improves the world in designers' or their clients terms. (2006, p. 31)

Krippendorff's influence can be seen here not only in distinguishing between forms of design, but also his prescience in anticipating a transdisciplinary design practice and the splintering over time into multiple subfields such as Industrial Design (ID), Interaction Design (IxD), User Experience (UX), Design Management (DM), and Service Design to name a few.

Among his many awards and publications, Krippendorff is perhaps best known for his book *The Semantic Turn: A new foundation for design* (2006) which is seen as a paradigm shift, an evolution from product semantics to semantics in relation to

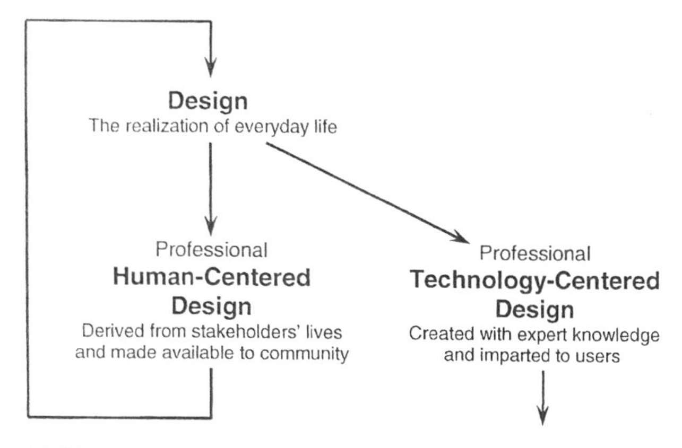

Fig. 6.4 Krippendorff's conceptualization of three types of design (2006, p. 32)

the meaning users ascribe to designed objects. Krippendorff recognized the various cross-disciplinary threads of influence that were converging during this period.

> "As may be imagined, the semantic turn that design is taking is correlated with several major intellectual, cultural, and philosophical shifts...It is also paved by radical changes in the social and technological environments in which design is now practiced." (Krippendorff, 2006, p. 13)

While a professor at the University of Pennsylvania, anthropologist Gregory Bateson (1904–1980) was Krippendorff's student in a course in Cybernetics, Language, and Culture.

Simply stated, semantics is a branch of linguistics that studies meaning and the nuance of meaning. For Krippendorff, the semantic turn in design was a shift away from the traditional functional and aesthetic concerns of design (i.e., product semantics) toward semantics as the study of communication and meaning, solidifying the role of the professional designer as working within a network of stakeholders. He argued that design should be in service to the user and that "the user" was not a faceless statistic, but rather a main stakeholder and participant in the design itself[5].

Krippendorff was also known for his work in *content analysis*[6], the study of documents and various forms of text, and *cybernetics*, the transdisciplinary study of "circular causal and feedback mechanisms in biological and social systems." (von Foerster et al., 1951) His interest in cybernetics brought him into contact with other anthropologists including Margaret Mead who perceived the role of cybernetics as "a form of cross-disciplinary thought which made it possible for members of many disciplines to communicate with each other easily in a language which all could understand." (Mead, 1968).

Although Krippendorff's influence on the evolution of design theory and practice cannot be overstated, for many if not most designers and design students the impact of his work goes unrecognized.

Donella Meadows: Thinking in Systems

Donella H. Meadows (1941–2001) was a pioneering environmental scientist, teacher, and writer known for her work in systems analysis and her leadership in sustainability education and advocacy. Meadows was the lead author of "The Limits to Growth"

[5] See Tamara Hale's excellent article on this topic "People Are Not Users." Hale argues that although ethnographic methods have been adopted in design practice, particularly in the technology industry, "the implementation of ethnographic methods has had less of an impact on the tendency to think of people primarily in relation to a specific product or service as "users", "customers" or "clients", which results in both a simplistic and individualistic view of human experiences." (2018, p. 163).

[6] Content analysis is another field in which Krippendorff made a significant contribution. Content analysis studies forms of communication including text, video, photographs, and audio. Identifying patterns in communication artifacts and assigning labels or "codes" to meaningful units of text are used to analyze meaning in content. He is credited with developing "Krippendorff's alpha", a statistical measure used to assess interrater reliability in content analysis.

(Meadows, 1972) a landmark report on the potential consequences of exponential economic and population growth within the finite limits of Earth's resources. This report was one of the earliest and most influential works in the field of sustainability and design. It sparked a global conversation about the need for sustainable development. The Club of Rome which was founded in 1968 gained attention after the 1972 publication of "The Limits to Growth". Membership in the Club of Rome included scientists, economists, businessmen, international high civil servants, heads of state, and former heads of state from around the world.

A 30-year update to "The Limits to Growth" was published in 2004. Since then, the concept of "degrowth" described as "a critique of ecological consequences of economic growth." (Kallis, 2018: 1) has entered the discourse. The concept of "degrowth" is closely related to the ideas presented by the Club of Rome, particularly in "The Limits to Growth." Degrowth is an economic and social movement that advocates for the downscaling of production and consumption. It suggests that reducing consumption and production can lead to a more sustainable and equitable world, addressing environmental degradation and resource scarcity. The movement emphasizes a shift away from the traditional measure of economic success as constant Gross Domestic Product (GDP) growth, arguing that perpetual growth is unsustainable on a finite planet. While the Club of Rome played a pivotal role in initiating discussions about sustainable development, the degrowth movement is broader and encompasses a variety of perspectives and strategies that extend beyond the Club's work. Degrowth is a diverse and evolving field, with contributions from economics, environmental science, and social theory, among others.

Meadows is perhaps best known for *Thinking in Systems: A Primer* (Meadows, 2008), a seminal work that introduces readers to the concept of systems thinking and its practical applications in understanding and addressing complex problems. Meadows provides a comprehensive overview of key principles and tools for system thinking highlighting nine key areas:

- Systems Thinking Fundamentals
- System Structure and Behavior
- Leverage Points
- Emergent Properties
- Resilience and Adaptation
- Ethics and Values
- Case Studies
- Interdisciplinary Approach
- Systems as Mental Models

Throughout the book, Meadows offers valuable insights into various real-world challenges. *Thinking in Systems* is a guide to developing a holistic and structured approach to problem-solving. It equips readers with the tools to analyze and intervene in complex systems, whether they are related to environmental sustainability, public policy, or any other area where understanding and managing complexity is crucial. Donella Meadows' work remains influential in fields ranging from environmental science, design, and business management, emphasizing the importance of systems thinking in a rapidly changing and seemingly chaotic world.

Meadows also developed the concept of "leverage points," places within a complex system where a small shift can produce significant changes in the system. In "Leverage Points: Places to Intervene in a System" (Meadows, 1999), a report for the Sustainability Institute, she outlined twelve places to intervene in a system in increasing order of effectiveness. Designers and practitioners from many fields continue to reference this report.

In addition to her research and writing, Meadows was a dedicated educator. She taught at Dartmouth College for many years and co-founded the Sustainability Institute (now the Donella Meadows Institute), which works to promote sustainable systems thinking and action. Her work has had a profound impact on fields such as environmental science, sustainability, and systems thinking. Her emphasis on the interconnectedness of systems and the need for sustainable development continues to influence scientists, policymakers, and activists today.

Enrico Manzini

Italian design strategist, Enzo Manzini, is a Professor of Design at Politecnico di Milano and the founder of the international network DESIS[7] (Design for Social Innovation toward Sustainability). He is recognized for his work on design for sustainability, with a particular emphasis on social innovation. Manzini's work is centered around the idea that design and social innovation can play a critical role in addressing complex societal challenges. He argues that by fostering local, collaborative, and environmentally friendly solutions, we can create a more sustainable and resilient society.

In his influential book *Design, When Everybody Designs: An Introduction to Design for Social Innovation* (2015), Manzini explores the relationship between design and social change. He argues that in a world where everyone has the ability to design (thanks to various tools and platforms), professional designers should act as facilitators, helping to initiate and sustain social innovation processes. Manzini's work on design for sustainability has been influential in shaping the field. He has highlighted the role of small, localized projects and everyday innovative practices in bringing about systemic change. He refers to these as "seeds of change" that, when networked together, can create a more sustainable society.

Through the DESIS NETWORK Manzini has helped to foster collaborations between design schools and other stakeholders to develop and promote design projects that address social and environmental issues. DESIS is now an international network with many active labs around the world.

---

[7] https://desisnetwork.org/

Enzo Manzini's work has been instrumental in promoting the idea of design as a tool for social innovation and sustainability. His emphasis on local, participatory, and collaborative approaches has influenced designers and design educators worldwide.

Lucy Suchman: A (Critical) Engagement Between Anthropology and Design

In "Anthropological Relocations and the Limits of Design" (2011), Lucy Suchman takes a critical anthropological approach to the engagement between anthropology and design, specifically, in the field of "'design science' that characterized interrelated disciplines of computer science, AI, management, and organization theory" (2011: 16). Drawing on 20 years of experience working in Silicon Valley (1979 to 1999), Suchman's critique is aimed at the hubris of technologists and designers. Rather than disruption she argues for.

> "…aspiring not to *massive change* [italics added] or discontinuous innovation but to modest interventions within ongoing, continually shifting and unfolding, landscapes of transformation. (2011: 16)

Suchman takes aim at the notion that design would save the world, a claim promoted in particular by graphic designer and design theorist Bruce Mau. In the excerpt above, she uses the term "massive change" in a direct reference to Mau's book *Massive Change* in which he asks, "Now that we can do anything, what will we do?" In this quote Mau seems to align himself with Papanek and others who warned of the consequences of unfettered growth. However, Suchman sees instead an inflated sense of self-assurance resulting from the increased attention to design's problem-solving potential.

> Our world now faces profound challenges, many brought on by innovation itself. Although optimism runs counter to the mood of the times, there are extraordinary new forces aligning around these great challenges, around the world. If you put together all that's going on at the edges of culture and technology, you get a widely unexpected view of the future. Massive Change charts this terrain. (2004, pp. 15-17).

Although design is widely recognized as playing a central role in domain of technology as well as in education, business, and medicine, Suchman challenges the this prevailing view of design by pointing to the hubris of Mau's statement.

Suchman is not the only one to raise the issue of design hubris. Anthropologist Tricia Wang (2021) argued in her *Fast Company* article that "The Most Popular Strategy in Design Thinking is BS: The 'How Might We' design prompt is insidious, and it's time to bury it." The article drew criticism from many designers. Wang's critique is that the current success and notoriety of design and designers has not been achieved without the adaptation of methods and theories from other disciplines and the appropriation of anthropology's signature methodology, ethnography. Design must become transdisciplinary, borrowing while recognizing its debt to the disciplines it draws from and aiming for "a responsible practice [that] is one characterized by humility rather than hubris." (Suchman, 2011: 16).

Arturo Escobar: Designs for the Pluriverse

Arturo Escobar is a Colombian-American and Professor Emeritus of Anthropology at the University of North Carolina at Chapel Hill whose work in the anthropology of development has focused on the impacts of "intensive globalization with its conspicuous collateral damage, including climate change, widespread extractivism, extensive conflict and social dislocation, and the inexpressible devastation of the Earth." (2017: ix) A main goal of *Designs for the Pluriverse* is Escobar's interest in exploring the potential for design to contribute to "communal forms of autonomy" that support the political struggles of Latin American indigenous, Afrodecendant, peasant, and marginalized urban groups as they defend not only their territories and resources, but also "their entire ways-of-being in the world."

Escobar takes the challenge posed by Papanek in *Designs for the Real World* (1973) beyond the design of technologically appropriate artifacts that are responsive to actual human needs. Instead, Escobar questions whether design's modernist tradition, based in a dependence on patriarchal capitalism, can transition toward relational modes of knowing, being, and doing. Along with the influences that inspired *Designs* including self-organization, autopoiesis, and complexity, Escobar identifies three threads that come together in his optimistic approach to design: his background in systems and engineering, his research on a "transition imagination exercise" for the Cauca River valley region (xiii), and his studies in ecology. The central role of imagination in transitioning from current unsustainable ways of life to envisioning possible futures based on "relational ways of knowing, being, and doing" recall the set of initial principles that cover the range of imagination identified by Pendleton-Jullian and Seely Brown (2018: 430–431) that were cited previously in this chapter. The notion of 'transitioning' introduces the next example of how designers are positioning for a post-technological world.

Transition Design: Terry Irwin, Gideon Kossoff, and Cameron Tonkinwise

Terry Irwin is a designer, academic, and researcher known for her pioneering work in the development of Transition Design, an area of design research, practice, and study that Irwin and her colleagues Gideon Kossoff and Cameron Tonkinwise introduced at Carnegie Mellon's School of Design. Transition Design is an approach that argues for transdisciplinary design-led societal transition toward more sustainable futures. Transition Design integrates knowledge from a range of disciplines and fields including sustainability and complexity science and calls for new ways of designing that are contextually aware, participatory, and transdisciplinary with a systems-oriented approach.

In Transition Design, the emphasis is on long-term thinking that incorporates historical understanding to trace how the current situation evolved and to identify inflection points within complex systems that nudge societies to transition to a more sustainable future. This involves imagining such a future, understanding how to bring it about, and developing steps that can lead us toward it. The approach acknowledges the interconnectedness of social, economic, political, and natural systems and proposes design as a catalyst for change within these systems.

Scott Boylston

Scott Boylston is an author, educator, practitioner, and sustainability advocate known for his work in sustainable design and design education. He is the founder of Emergent Structures, a non-profit organization that focuses on building structures using reclaimed and repurposed materials to address social and environmental issues. Boylston's work centers around the concept of sustainable design, which involves designing products, systems, and processes that minimize negative impacts on the environment and society while promoting social and economic well-being (2019). He emphasizes the importance of considering the entire lifecycle of a design, from sourcing materials to end-of-life disposal.

In his writings and teaching, Boylston encourages designers to think critically about their role in creating a more sustainable future. He promotes the integration of sustainability principles into design education and emphasizes the need for designers to be socially and environmentally responsible in their practice.

Boylston's work emphasizes the power of design to address pressing global challenges, such as climate change, resource depletion, and social inequality. His approach is multidisciplinary, involving collaborations with various stakeholders and communities to create meaningful and sustainable solutions.

In a recent social media post[8], Boylston cites the work of Norwegian philosopher Arne Naess who proposed a *deep ecology* that he describes as

> one that viewed humankind as a single species among many that exists not at the top of any imagined hierarchy, but as an equally interdependent one within a networked *holarchy* [italics added]. The goal of the deep ecology movement since has been to shift society's values, attitudes, and lifestyles toward celebrating and safeguarding this vibrant diversity for its intrinsic value, and to instill '*a profound awareness of the difference between bigness and greatness.*' (Boylston, 2023)

[8] https://www.linkedin.com/pulse/exploring-nature-inclusive-design-principles-scott-boylston%3FtrackingId=Ga6sPEWyTb6oZbSCujORZQ%253D%253D/?trackingId=Ga6sPEWyTb6oZbSCujORZQ%3D%3D.

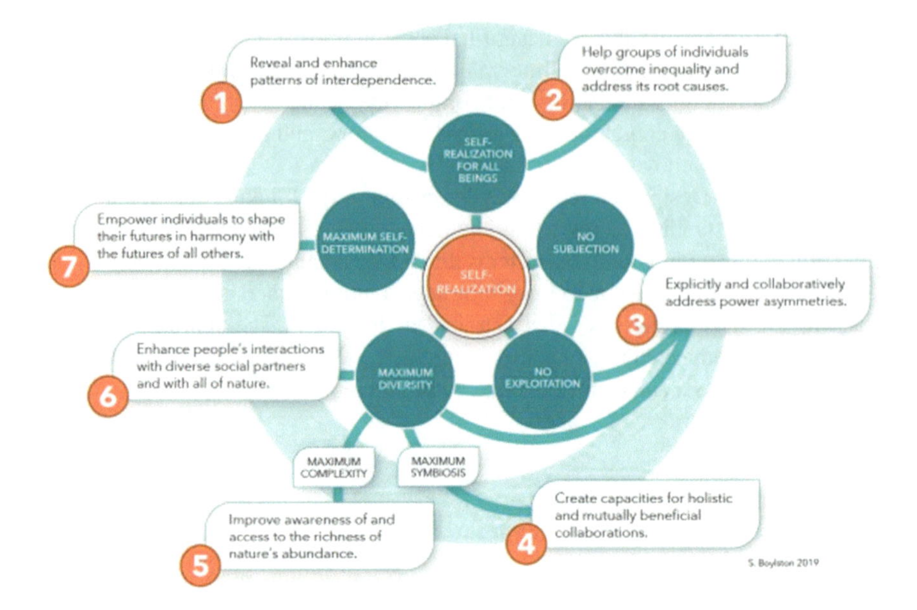

Fig. 6.5 Exploring Nature Inclusive Design Principles (Boylston, 2023)

As a co-author of the Design for Sustainability program at the Savannah College of Art and Design (SCAD) and a designer with over 45 years of professional experience in various design and non-design arenas, Boylston has consistently articulated ethical principles that guide his work. He has helped many students do the same. His advice for those interested in building a design career that broadens their human-centered design scope is to encompass all of nature and to craft a personal "ecosophy" to guide their professional work (Boylston, 2023) (Fig. 6.5).

Manuel Lima

Lima introduces himself as a person who is interested in "inspiring creative vision through passion and curiosity."[9] As a designer, lecturer, researcher, and author Lima is known for his work on the visualization of complex networks. His books include *Visual Complexity: Mapping Patterns of Information* (2011), *The Book of Trees: Visualizing Branches of Knowledge* (2014), and most recently *The New Designer: Rejecting Myths, Embracing Change* (2023) have established him as an influential voice in the field of information visualization. Lima is a strong proponent of the notion that designers have a substantial effect on the world that is accompanied by personal,

[9] Manuel Lima on LinkedIn "Inspiring creative vision through passion and curiosity. Follow for insights on Art, Design, UX, History, Technology, Data Visualization and Visual Culture." https://www.linkedin.com/in/mslima/ Accessed July 20, 2023.

societal, cultural, and environmental responsibility. Shaking off the dominant view that design is restricted to aesthetics, Lima rejects the idea that designers can pass accountability to the entities that compensate them for their work. He advocates that designers can be a "force for good", focusing on design that is ecologically responsible and serves humanity in all its variations "in a society increasingly driven by metrics, algorithms, and profit" (2023).

Illustrating the transdisciplinary trajectory in design (Miller et al., 2022), Lima's work extends beyond design and draws from fields including ethics, economics, psychology, and ecology. He epitomizes a new paradigm in design that aligns with our concept of a post-technological perspective. However, he is not the first designer to propose an outsized role for design, claiming that "Design is the Answer", the title of one of the chapters in *The New Designer*. As noted in the critiques of Lucy Suchman and Tricia Wang, the tendency to amplify the role of design in solving society's most pressing problems is a common thread running through the discourse of design.

Technology-Enhanced Imagination and Speculative Design

Technology is unlikely to outpace the human imagination, or so we wish to think. Yet when technology enables the imagination, it is hard not to argue that the outcomes can be unbelievable, stunning, and even alarming. The case described above illustrates how technology plays an enabling role in creating the conditions to transform the fictional. Hannah Kim, assistant professor of philosophy and associate editor at the Stanford Encyclopedia of Philosophy, wrote "Kondo and Hatsune's relationship became much more serious after he purchased a hologram machine that allowed them to converse." Generative A.I. is the latest installment in the steady stream of game-changing technologies to raise fears that we have somehow crossed a line. There is, in fact, a term for the sense that we are in uncomfortable and possibly dangerous territory. The term "uncanny valley", introduced by Japanese roboticist Mori Masahiro[10] in 1970 (Mori, 1970; Mori et al., 2012), is used in robotics, artificial intelligence, and human–robot interaction to describe feelings of revulsion, eeriness, or a general negative emotional response to a robot or chatbot that becomes increasingly empathetic or aggressive, or one that too closely resembles human appearance. Hannah Kim notes that the human-like responses of interactive chatbots like Microsoft's Sydney allow us to engage in "interactive fiction". We care about them as we care about fictional characters in a great novel, but we can also interact with them, and those interactions can produce outputs that are uncomfortable, alarming, or even harmful.

Design plays an active role in the robotics example. Masahiro Mori noted the mathematical term *monotonically increasing function* that describes a relation in which $y = f(x)$ increases continually with the variable x, e.g., when we put pressure

[10] We follow the Japanese practice of placing the family (Mori) before the given name (Masahiro).

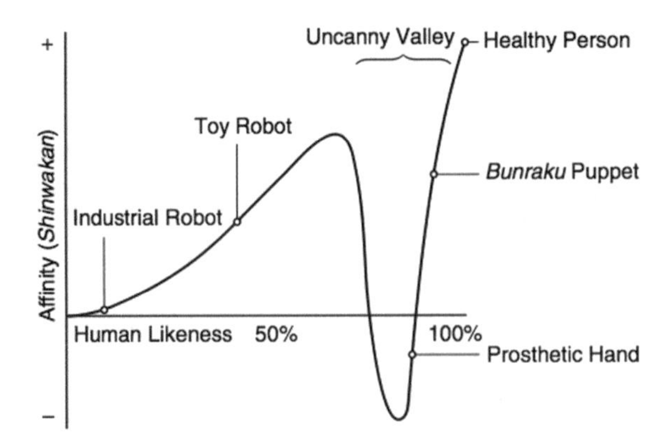

Fig. 6.6 The graph depicts the uncanny valley, the proposed relation between the human likeness of an entity, and the perceiver's affinity for it. [Translators' note: Bunraku is a traditional Japanese form of musical puppet theater dating to the seventeenth century. The puppets range in size but are typically a meter in height, dressed in elaborate costumes, and controlled by three puppeteers obscured only by their black robes.] From Mori et al. (2012) (Fig. 6.7).

on the accelerator of a vehicle it goes faster. Yet we know that applying continuous effort can result in an unwanted outcome. More becomes too much. In designing robots to appear more human-like Mori states "Our affinity for them increases until we come to a valley, which I call the *uncanny valley*." (2012: 98) Because their *design* is functional industrial robots, or *co-bots* (Biton et al., 2022) collaborative robots that are a driving force on factory floors, do not fall into the uncanny valley. However, the designer of a toy robot, or a robot designed to provide a service (i.e., *carebots*) will focus on more human-like features that provoke a greater affinity. Materials can be used to design a human-like look and feel such as skin texture. When the designer achieves a likeness that *looks* to be human, people tend to accept a robotic device, such as a hand. However, once they realize that the hand is a prosthetic hand, perhaps by touching it, the sense of eeriness sets in. "The hand becomes uncanny." Adding movement, such as the grip of a handshake, makes the sensation even more disturbing (Fig. 6.6).

"Escape by Design"

Mori addresses the challenge of designing and building prosthetics and robots that have functionality and human-like features and elicit affinity with humans. He predicts "that it is possible to create a safe level of affinity by deliberately pursuing a nonhuman design." (Mori, 2012, p. 100) Mori gives examples such as the model of a human hand sculpted in wood by a woodcarver that sculpts statues of Buddhas. He notes "The fingers bend freely at the joints. The hand lacks fingerprints, and it retains the natural color of the wood, but its roundness and beautiful curves do not

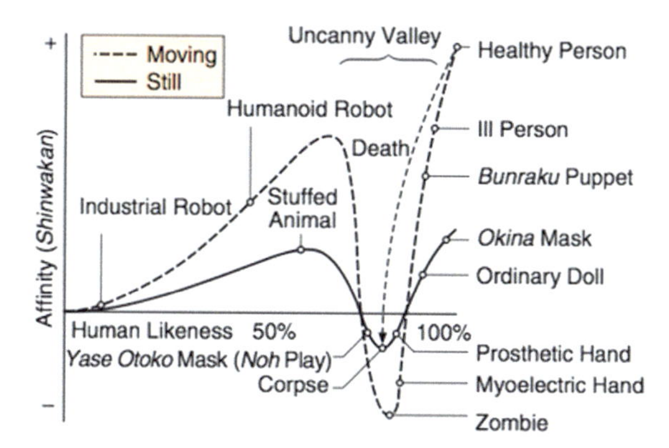

Fig. 6.7 The presence of movement steepens the slopes of the uncanny valley. The dotted line's path represents the sudden death of a healthy person. [Translators' note: Noh is a traditional Japanese form of musical theater dating to the fourteenth century in which actors commonly wear masks. The *yase otoko* mask bears the face of an emaciated man, and represents a ghost from hell. The *okina* mask represents an old man.] From Mori et al. (2012)

elicit any eerie sensation. Perhaps this wooden hand could also serve as a reference for design."

Our focus on Mori's concept of the uncanny valley has a purpose, which is to set up a design related question. Anthropologist Trisha Wang called out what she characterized designers' go-to question "How might we…?" as BS (Wang, 2021). Although the Fast Company article drew some flack, the prevalence of the term to the extent that it became an acronym HMW made Wang's point. Beyond the debate and acknowledging that "we" has no reference point aside from the people or person asking the question and does not consider the implications for others, the term underscores a design heuristic. This is where abductive logic kicks in, questions begin to surface, and consideration is given as to how to work with constraints. This is the point at which imagination is accessed with the deceptively simple question *"What if?"*.

What if Mori's concept of the uncanny valley were used to inspire the development of a framework to guide the design in the application of technology?

References

Akrich, M. (1992). The De-scription of Technical Objects. In Bijker, W, Law, & J (Eds.), *Shaping Technology/Building Society. Studies in Sociotechnical Change* (pp. 205–224). MIT Press. https://shs.hal.science/halshs-00081744

Becker, G. S. (1976). *The Economic Approach to Human Behavior*. University of Chicago Press.

Biton, A., Shoval, S., & Lerman, Y. (2022). The Use of Cobots for Disabled and Older Adults. In *14th IFAC Workshop on Intelligent Manufacturing Systems IMS 2022.*

Bourdieu, P. (1977). *Outline of a theory of practice*. Cambridge University Press.

Boylston, S. (2019). *Designing with society: A capabilities approach to design, systems thinking, and social innovation.* Taylor & Francis.
Boylston, S. (2023). *Exploring Nature Inclusive Design Principles* LinkedIn.
Brown, T. (2009). *Change by design.* HarperCollins Publishers.
Buchanan, R. (1992). Wicked problems in design thinking. *Design Issues, 8*(2), 5–21. https://doi.org/10.2307/1511637
Bush, V. (1925/2021). *Science, the endless frontier.* Princeton University Press.
Butler, J. (1990). *Gender trouble: Feminism and the subversion of identity.* Routledge.
Chia, R. C. H., & Holt, R. (2009/2011). *Strategy without design: The silent efficacy of indirect action.* Cambridge University Press.
Clarke, A. J. (Ed.). (2018). *Design anthropology: object cultures in transition.* Bloomsbury Academic.
De Mozota, B. B., & Valade-Amland, S. (2020). *Design: A business case.* Business Expert Press.
Giddens, A. (1984). *The Constitution of Society: Outline of the Theory of Structuration.* University of California Press.
Hale, T. (2018). People Are Not Users. *Journal of Business Anthropology, 7*(2), 163–183.
Halse, J. (2013). Ethnographies of the Possible. In T. O. Wendy Gunn, and Rachel Charlotte Smith (Ed.), *Design anthropology: Theory and practice.* Bloomsbury.
Illich, I. (1973). *Tools for conviviality* (Vol. 47). Harper & Row.
Kallis, G. (2018). *Degrowth.* Agenda Publishing.
Kant, I. (1788). *Critique of practical reason.* Hackett Publishing Company.
Kim, H. H. (2023). A.I. is a fiction: Stop freaking out when chatbots say they're in love or make disturbing threats. Just treat them like Pinocchio. *Wired*, (31.09).
Krippendorff, K. (2006). *The semantic turn: A new foundation for design.* Routledge.
Kumar, V. (2013). *101 design methods: A structure approach for driving innovation in your organization.* John Wiley & Sons.
Latour, B. (2006). Mixing humans and nonhumans together: The sociology of a door-closer. In W. B. William, A. H. Jonathan, & J. S. Sadar (Eds.), *Rethinking technology: A reader in architectural theory* (p. 16). Routledge.
Law, J. (1987). Technology and heterogeneous engineering: The case of portuguese expansion. In W. E. Bijker, T. P. Hughes, & T. Pinch (Eds.), *The social construction of facts and artifacts: new directions in the sociology and history of technology.* MIT Press. (1987) (Reprinted from 1999)
Libet, B. (1999). Do we have free will? *Journal of Consciousness Studies, 6*(8–9), 47–57.
Liedtka, J., & Ogilvie, T. (2011). *Designing for growth: A design thinking toolkit for managers.* Columbia University Press.
Lima, M. (2011). *Visual complexity: Mapping patterns of information.* Princeton Architectural Press.
Lima, M. (2014). *The book of trees: Visualizing branches of knowledge.* Princeton Architectural Press.
Lima, M. (2023). *The new designer: Rejecting myths.* MIT Press.
Mau, B. (2004). *Massive change.* Phaidon Press.
Mead, M. (1968). The cybernetics of cybernetics. In H. von Foerster, J. D. White, L. J. Peterson, & J. K. Russell (Eds.), *Purposive systems* (pp. 1–11). Spartan Books.
Meadows, D. H. (1972). *The limits to growth: A report for the club of Rome's project on the predicament of mankind.*
Meadows, D. (1999). *Leverage points: Places to intervene in a system.*
Meadows, D. (2008). *Thinking in systems: A primer.* Chelsea Green Publishing.
Miller, C. Z., Palsikar, S., & Spears, J. M. (2022). Evolving praxis in design management: The transdisciplinary trajectory. *DMI:Journal, 17*(1).
Mori, M. (1970). Bukimi no tani [The uncanny valley]. *Energy, 7*, 33.
Mori, M., MacDorman, K. F., & N. Kageki. (2012). The uncanny valley [From the Field]. *IEEE Robotics & Automation Magazine, 19*(2), 98–100. https://doi.org/10.1109/MRA.2012.2192811

Ortner, S. (2006). Power and projects: Reflections on agency. In *Anthropology and Social Theory: Culture, Power, and the Acting Subject* (pp. 129–154). Duke University Press. https://doi.org/10.1515/9780822388456-008

Papanek, V. (1973). *Design for the real world: human ecology and social change.* Bantam Books.

Pendleton-Jullian, A. M., & Brown, J. S. (2018). *Design unbound: Designing for emergence in a white water world* (Vol. 1). MIT Press.

Reznich, C. (2017, July 17). 1973 Horst Rittel.

Rittel, H., & Webber, M. M. (1973). Dilemmas in a general theory of planning. *Policy Sciences, 4*(2), 155–169.

Sanders, E. B.-N., & Stappers, P. J. (2014). *Convivial toolbox: Generative research for the front end of design.* Colophon.

Sartre, J.-P. (1943). *Being and nothingness.* Washington Square Press.

Suchman, L. (2011). Anthropological relocations and the limits of design. *Annual Review of Anthropology, 40*, 1–18.

Vetter, A. (2017). The matrix of convivial technology—Assessing technologies for degrowth. *The Journal of Cleaner Production, 197*.

von Foerster, H., Mead, M., & Teuber, H. L. (1951). *Cybernetics: Circular causal and feedback mechanisms in biological and social systems.* Transactions of the seventh conference. Macy's Cybernetics Conference, New York.

Wang, T. (2021, June 2021). The most popular strategy in design thinking is BS *Fast Company.*

Zachary, G. P. (1977). *Endless frontier: Vannevar bush.* Free Press.

Chapter 7
Narrative of the Machine

> *"Things are in the saddle, and ride mankind."*
> *Ralph W. Emerson, Ode Inscribed to William Ellery Channing, 1847*

Abstract In this chapter, we develop a narrative account of technology, building on techno-celebrants such as Lewis Mumford, techno-skeptics such as Jacques Ellul, and critical technology theory from Martin Heidegger, Jurgen Habermas, and Andrew Feenberg. It points up the fundamentally *irrational* attachment to technology that is resolved by seeing technology as a *symbolic,* rather than a rational construction. We examine the common core of these narratives or myths, to determine what they reveal regarding the *character* of technology in contemporary life. This character, sometimes noble, sometimes villainous, shapes contemporary life in the technological society no less than the institutions of government or religion. Narratives, we argue, are no less important than mechanisms for determining that shape of the technological society.

When we think of technology we immediately think of tools and artifacts, and then the uses to which these are put, whether the computations of a spreadsheet or the soaring of an airplane. As our reader is undoubtedly aware, technology is far more complicated than simply tools, artifacts, and standards. Technology, in fact, embraces numerous *shared narratives* of a technological society. At the top of these narratives, of course, is man's mastery over Nature, a narrative that has played out in Western society for thousands of years. Tolstoy's 1878 novel, *Anna Karenina* has most of its action either on trains or in the St. Petersburg train station, and against the backdrop of the liberalizing reforms of Czar Nicholas Alexander II the technology constructs a momentum of change that came crashing down with the Russian Revolution in 1917. The train journey is in fact a *liminal space,* "betwixt and between," suggesting not simply a transition between St. Petersburg and Moscow, but a transition away from the czarist era that was drawing to a close. Liminality, in the crevices between institutional orders, is a dynamic of social evolution. *The Machine in the Garden* (Marx, 1964), for but one of many examples, describes the collision between industry

A. Batteau and C. Z. Miller, *Tools, Totems, and Totalities*,
https://doi.org/10.1007/978-981-97-8708-1_7

and unspoiled Nature. Other traditions, whether the imperial ambitions of Russia after the Industrial Revolution, or Asian stories of the Kingdom of Heaven also construct narratives around human-built tools.

Storytellers are no less important to the technological society than engineers and factory workers. Storytellers allow us to imagine the world of abundance and ease that technology promises. Narrative is no less an institutional foundation of the technological society than are roads and bridges, and understanding the stories we tell about technology is no less important than understanding its thermodynamics and cybernetic properties. In this chapter, we examine the narrative aspects of technology to better understand the narratives of a world beyond the hegemonic technology as we know it today.

In many of these narratives, technologies play a central role, sufficiently so that we can question what is the *character* of technology: strong v. weak, masculine v. feminine, wise v. foolish, parochial v. worldly? Information and social media, for but one of many examples, beguiles the users into its own realities, most notably virtual (see Chap. 11 on the "metaverse"). Similarly, the railway, with its loud, clattering noise of the steam engine from the last century impressed everyone with the machine's mastery over forms of human mobility. The radio and television persuaded millions of viewers that they had a *connection* with distant consociates on other continents. These realities, connections, and transitions are the bone and muscle of modern society. Lewis Mumford, in The Myth of the Machine (1966), describes the "megamachine" as an all-pervasive enlistment of human capabilities toward impersonal ends. For Mumford, modern urban society orchestrates industry and architecture and technics into a large, complex, humanity-dominating machine. A *myth*, we should make clear, is not simply a fabulation, but a *narrative* that is told to capture some larger Truth, whether the myth of the Garden of Eden (and the descent of Man from a golden age) or the myth of the Russian Revolution and its overthrow of czarist tyranny. Myths can be heroic or tragic, or sometimes both. Julius Caesar, the first Roman emperor and conqueror of Gaul, was also a tragic figure who met a tragic death on the Ides of March in 44 BC. Myths are the stories that knit a community together, both among members of the community and with the larger cosmos. Just as the story of Romulus and Remus knit together the early Romans and the landscape, and has been retold for thousands of years since, every nation uses its historical narratives, sometimes literally factual but more often embellished, to recount its greatness. A recounting of national myths of greatness would fill an entire library. It should suffice, for the moment, to note that *nations*, as contrasted to other aggregations, are built around heroic myths, whether that of the Founding Fathers or the Magna Carta. Continents do not have myths, nor do census tracts nor seacoasts nor soup kitchens.

Christopher Lasch, showed how Mumford's "myth of the machine" drew on major Western thinkers, from Marx to Weber, describing a "mechanistic plan of life" as the apogee of capitalism. In Lasch's words, "Whatever its liberating possibilities, Mumford insists, modern technology has created new forms of enslavement because it has grown up as the servant of national self-aggrandizement, war, and the insatiable appetite for profits" (Lasch, 1992, p. 104). The human filing case, reducing individuals to consumers dependent on the market, is the logical outcome of the myth of the machine.

In this chapter, we examine the common core of these narratives or myths, to determine what they reveal regarding the *character* of technology in contemporary life. This character, sometimes noble, sometimes villainous, shapes contemporary life in the technological society no less than the institutions of government or religion. As we unpack the myths of the machine, we can lay the foundation for understanding the *character* of contemporary technological society. Narratives, we argue, are no less important than mechanisms for determining that shape of the technological society.

The Myths of the Machine

A classic narrative of the myth of the machine is a 1922 silent movie by Fritz Lang, *Metropolis.* This movie presents a dystopian view of a city as a megamachine in which the workers toil underground in service of the master, Freder. Some of the images associated with the movie such as the *machinemensch* have become iconic of an earlier era in silent movies (Fig. 7.1).

Metropolis epitomizes the "megamachine" that nearly a half-century later Lewis Mumford, in *The Myth of the Machine* described as embracing all modern humanity. Roughly contemporary with Metropolis, Franz Kafka's *The Castle* (1926) provided a chilling portrait of life under the control of a distant bureaucracy. The distant

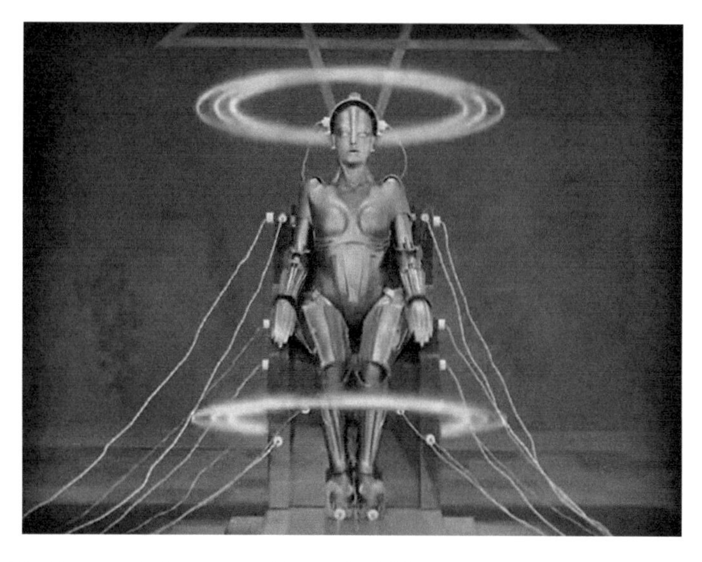

Fig. 7.1 Machinemensch (from Lang, *Metropolis,* 1922)

mysterious mayor, residing in a castle, controls the villagers, even if they do not know his name or what he does.

Another story that has resonated throughout the technological era is the conquest of distant continents and ultimately outer space. Initially created by Marco Polo in China and later exemplified in Jules Verne's *de la Terre a la Lune* (1865), which have served as inspiration not only for the space race of the 1960s (itself a heroic story) but also for fantasies of the conquest of space ever since. As we make clear in Chap. 11, "The Emperors' New Clothes," dreams of the conquest of space, whether missions to Mars or voyages beyond the solar system, are fantastical, requiring sustained investment over decades if not centuries, at a time when humanity's attention span is becoming notoriously short.

Perhaps a final narrative of heroic technology comes from the Battle of Britain, in which the Royal Air Force, assisted by the new technology of radar, overcame the numerically superior Luftwaffe. This story has been retold many times in cinema and history and has inspired many others to view technology as a key to military superiority. Through the past 75 years, the world's biggest investor in technology has been the American military, at times yielding many civilian benefits, but also making the world a more dangerous place. This view of military superiority can be delusional: when 19 terrorists, armed with box-cutters on September 11, 2001, brought a sophisticated technological power to its knees, it dramatically illustrated the limits of technology. Never before in the field of human conflict have so many been so terrorized by so few.

In all of these stories we can see technology as defining a *character*, at times heroic and masculine, sometimes villainous, but never weak or feminine. The *character* of technology comes from the part that technology plays in technological dramas, and is, in fact, instrumental in determining the technology's consequences, no less than functionality, force, mass, and acceleration. The character of technology includes man's transcending earthly limitations, reaching into space, dominating subordinate populations or exhibiting masculinity through strength. The character of technology is never feminine or childish, but instead heroic by saving civilization (or, inversely destroying it).

Character and narrative, as we will elaborate in the next chapter, are central for defining who we are and where we fit into the universe. Narratives hark back to ancient stories, sometimes tragic, sometimes heroic, and more interestingly, sometimes a combination of both. They are less a step-by-step accounting of events and more a drawing on ancient stories, whether the fall of man in the Garden of Eden or the conquest of the frontier in American mythology. The conquest of the frontier, perhaps *the* central narrative of America (along with the War for Independence), enlists some major archetypes, themes, and characters, whether the conquest of humanity over nature (later immortalized as Nature) or the frontiersman Daniel Boone and his later epigone, the cowboy. Character in drama, as Stanislavski (1950) makes clear, is constructed, not simply emergent, built from gestures, accentuation, intonation, and other elements. A good actor knows how to *build* a character and charm (or terrorize) his or her audience.

Character, we must make clear, is not simply a part of a dramatic script, but rather a fundamental building block of a civilization, uniting both social role (spouse, boss, employee, citizen, scholar, political leader) with some of the basic narratives of the civilization. We can judge political leaders, citizens, bosses, and employees in terms of how well they fulfill their destiny. Some civilizations have crashed because their central characters, whether Napoleon or Hitler or Nero, were fundamentally flawed. Their character defects were enlarged into the compass of their entire civilization; others have persevered and become models to the world, whether Winston Churchill or Franklin D. Roosevelt. A compromised character can never make his country great.

In Lewis Mumford's *The Myth of the Machine* (1967), both communism and capitalism offered a "mechanistic plan of life" far divorced from the sacred attachments of earlier years. For Mumford, the technological determinism of Marx and other writers would be replaced by a view that saw social patterns reproducing themselves at multiple levels. For Mumford, modern technology creates new forms of enslavement in the apparatus of commodity production. The city is reduced to a "human filing case" and the megamachine, a view prefigured in *Metropolis*, offers a new perspective of humanity. In the century since *Metropolis* and *The Castle,* technological improvements, whether facial-recognition systems or location-tracking cell phones, have only improved technology's surveillance of humanity. (See below, Chap. 10 on China's "social credit system.").

An ironic aspect of the myth of the machine is how humanity often *accepts* this subordinate role, perhaps because we find it more meaningful than unconstrained freedom. A scripted, meaningful role is spiritually more satisfying than unscripted freedom, and in fact can be seen in the enthusiastic followers of many dictators, whether Hitler, Mussolini, many Latin American dictators, or more recent authoritarian politicians. Autocrats thrive on the enthusiasm of their followers.

Homo Ludens

One of the stories that we tell about humanity is that humanity's species, *homo sapiens* (thinking man), who succeeded *homo habilis* (capable man) on the evolutionary scale, is advanced rather than primitive. In the standard narrative, *homo sapiens* emerged with the Neolithic Age (New Stone Age) where humanity evolved from foraging in scattered settlements to agriculture and urban settlements. With the neolithic, increasingly tools and instrumentalities came to play a role in human evolution.

An alternative view is presented by the Dutch cultural historian Johan Huizinga, in his book *Homo Ludens* (Huizinga, 1938) (playful man), suggesting that play is an essential part of humanity. *Play* is the unconstrainted experimentation of relationships (both social and physical), and in fact this element of experimentation and exploration is a key to evolution. By *exploring* new possibilities species discover new opportunities, new resources are discovered, and new adaptive possibilities emerge. These adaptive possibilities include new possibilities of relationships and new forms

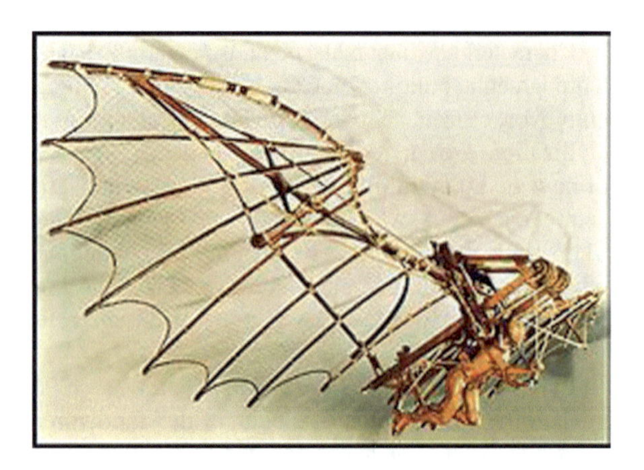

Fig. 7.2 Leonardo Da Vinci's flying machine

of authority, whether the Athenian experiments with rule by the people or the Sumerian invention of bureaucracy and irrigation. Even thousands of years later we imagine cities, architectural, and industrial masterpieces, as the epitome of civilizations.

The characteristics of play, for Huizinga, include freedom, unreality, liminality, order, and a disconnect from material interest. This freedom enables humanity to explore new social forms and hence to evolve. The printing press, e.g., emerged from experimentation with moveable type, and the computer emerged from playing with bits of code. Exploration and experimentation, from the perspective of cultural evolution, are no less important than the exploitation of resources in the immediate environment. The sky above, e.g., has existed for millions of years, and Greek fables of Icarus who died from flying too close to the sun pointed to a tragic fate, but only from the Renaissance did humanity imagine exploring it, first with Leonardo's flying machine, and later with flying devices that enlisted scientific findings of force and mass and acceleration, culminating in the Wright brothers' first flight in 1903 (Fig. 7.2).

Over the next 120 years, the airplane has grown to play an immense role in industry, warfare, business, and leisure, and has knit together the world, dissolving national borders.

Narratives of Technology

Some of the notable stories that have been told about technology include Kafka's *The Castle,* Upton Sinclair's *The Jungle* (1906), and Jules Verne's *Twenty Thousand Leagues Under the Sea*. Each of these has served as catalysts or templates for emergent institutions, social forms and technologies. *The Jungle*, e.g., described the meatpacking industry in Chicago at the beginning of the twentieth century, and led to the creation of the Food and Drug Administration (FDA) in 1906 to ensure food safety at a time when fewer and fewer Americans were living on farms and raising their own food. Food is part of how a people define themselves, and the industrialization of the food supply in the twentieth century is part of the industrialization of all aspects of domestic life. *The Jungle* presented a dystopian image of the American food supply, resulting in a changed relationship with the federal government.

Similarly, Franz Kafka's *The Castle* (1926) written in Weimar Germany immediately before the Nazi takeover describes a society dominated by a massive bureaucracy, describes alienation, submission, and isolation, and an unresponsive bureaucracy in which a dominating and inaccessible power resides high in a castle on a mountain overlooking society. This theme has been revisited ever since as the powers of governments and corporate bureaucracies become ever more immense. The Internal Revenue Service (IRS), America's tax collector, is often held up as a distant oppressive machine, at least by those who think they are paying too much in taxes. Frank Norris's novel, *The Octopus* (1901), describes the conflict between ranchers and farmers and the Southern Pacific Railway.

We could add further examples of narratives of the machine, from stories of Roman chariots to fables of the automobile. In all these the machine is a human creation detached from humanity, at times affecting human communities in unanticipated ways. Bureaucracy, in many portrayals, is also a machine. Political *machines,* whether municipal or national, have gears and pulleys that grind citizens and constituents into votes and tax payments, a largesse harvested by the machine, whether Richard Daley's Chicago or Tammany's and Boss Tweed's New York. The entire conception of a political machine is roughly coincident with the industrialization of the economy, and the imagery of a machine which negates humanity, denying both free will and sacrality, has become a significant part of the contemporary imaginary.

The Characters We Keep

Technological dramas are a well-established fact within the developmental cycle of any technology, from the extension of new capabilities (at times, but not always, revealed by scientific findings) through ascent to maturity to decline. The steam engine, e.g., invented by James Watt in 1712, was a central part of the railways that knit together continents in the nineteenth and twentieth centuries. Steam locomotives are today in decline, being replaced by diesel and electric motors. One of the great technological dramas today is the contest over fossil fuels, with antagonists concerned over global warming vs. those whose states and regions are dependent on a fossil fuel economy. This drama pits the survival of the planet (perhaps the ultimate existential threat) against the wealth of industrialists, a story that is ongoing.

The concept of technological drama (Pfaffenberger, 1992), which we discuss in Chap. 11, looks at humans and groups as the actors in the drama, through processes of confrontation, escalation, and resolution. We might also look at the artifacts, standards, and representations as having both agency and character, or more accurately, in conjunction with their *design,* which we explore in Chaps. 5 and 6, and use constituencies. That is to say that the constituencies and the artifacts form a team that would be inconceivable without its constituent elements, just as aviators need their airplanes (and vice versa), the *character* of Barney Oldfield was formed in conjunction with his race car, Wyatt Earp with his six-shooter, and Bill Gates with his desktop computer. We can analyze the *character* of these constitutive parts as characters in a larger drama that takes on a new aspect in an industrial and technological society.

Having posed this question, we should be asking what are the larger dramas of a technological society? What are the stories that we tell that inexplicably involve technologies? At the top of the list, of course, are international contests, whether the space race between the US and Russia in the 1960s, or the twentieth century wars in which technologies played a key role. The mercantile era, lasting roughly from the fifteenth to the eighteenth centuries, used seagoing power, trade and merchant vessels, including warships to build multinational empires. In the previous eras, the weapons of war were the foundation of empires. The Industrial Revolution was also a technological drama, with all of the constituent elements that Pfaffenberger identifies regularization, adjustment, and reconstitution. Factories represent a *regularization* of production (central location, product, and process supervision) that domestic production lacks. A*djustment* of workers to factory regimentation was an innovation of the Industrial Revolution. The *reconstitution* of family and kindred was the closing act of the drama. With this drama, the world was changed utterly, beginning in Europe and later in the remainder of the world. The Industrial Revolution, a scaling up of production, transformed the world from one of imperial domination to one of cores and peripheries, with the first core region being northwestern Europe.

Immanuel Wallerstein, in *The Modern World System* (1988), argued that in the twentieth century the new architecture of the world consisted less of empires and colonies and more of "core" and "peripheral" states. Northwestern Europe was a core region, as was most of North America. Africa and Latin America were the periphery, with a "semi-periphery" of southern Europe and Mexico and Brazil in between. Core states had diversified, inclusive economies, and create self-generating growth, while peripheral states had extractive economies (based on mining and natural resources) and were dependent on the core region. Peripheral economies were divided, whereas the core, as exemplified by the European Union, was integrated (Fig. 7.3).

Characterizing the development of industrial production on a large scale draws on imperial archetypes and dynastic rise whether in Byzantium or the Middle Kingdom). Core nations exert their dominance over the periphery primarily through technological and industrial innovation. Similarly, in the twentieth century, the rise of the technological society, as anyone who was sentient through it remembers, was a dramatic event with new products, new perspectives, and new social arrangements between the core nations of Europe, China, and Japan and the peripheries of Africa and Latin America.

The division of the world between technological core and technological periphery is a recent dynamic. *Technological* peripheries, whether in sub-Saharan Africa or parts of Latin America, are frozen in their subordinate positions due to a lack of social and human capital, much as peripheral *economies* are trapped by a lack of finance capital. The instruments of core nations, whether transportation devices or

Fig. 7.3 A model of a core-periphery system as used in dependency or world-systems theory. Such systems were proposed by Andre Gunder Frank and Immanuel Wallerstein, among others

information technologies or medical technologies and medical institutions, solidify the dominance of core over periphery far greater than the Titans of finance.

In all of this, we see a performative aspect of technology, a *staging* of dramatic contests, in which the users and spectators are enthralled (or horrified) by the technological drama. Technological dramas are no less important for knitting together the human community than the Dionysian dramas of ancient Athens or the Noh plays of Japan. Every major civilization has used theater and drama to make a statement about "this is who we are."

Designers have a role in this evolution, as noted in the previous chapter's discussion of Victor Papanek. Papanek considered that industrial and advertising design, central to today's economy, subversive to the character of contemporary civilization. *Advertising* is some of the main currency of today's society, yet all recognize that it should not to be taken too seriously. When we roll up our sleeves to design a civilization beyond technological hegemony, we have a choice to make of its character, of its authenticity or dishonesty, its fulfillment of human purposes, or enslavement to the machine.

Drama and Identity

The stories we tell, the manner in which we *dramatize* their consequences, are a central part of establishing our identity, whether as heroic conquerors of the wilderness or fallen children of a merciless god. Humanity, of course, has been telling these stories for millennia, but in the modern era, a new character, named technology, has joined the narrative. This character, as illustrated above, is at times a friend, and at other times a malign influence. Far more than simply a *tool* (an extension of human capability), technology is a *character*, an actor within networks of human and nonhuman actors in larger-than-life dramas.

Having given our technologies an identity, we can ask what stories they tell and in what manner they help build our character or our identity as members of society. There are, of course, dozens if not hundreds of archetypal narratives which societies draw upon to establish their identity. An early example of this would be Thucydides account of the Peloponnesian War, in which the funeral oration of Pericles has served as a model for the commemoration of heroes. Echoed in Lincoln's Gettysburg Address it began with an acknowledgement of the heroes and their sacrifices. A similar theme, in Shakespeare's Henry V, at the battle of Agincourt, where Henry celebrated the "band of brothers" who overcame a numerically superior French army on St. Crispin's Day in 1415. And finally, the heroic tale of Charles Lindbergh's flight from New York to Paris created a new national hero who vaulted from obscurity to international celebrity in the space of three days in 1921. Lindbergh was celebrated for his feat upon his return from Paris.

In America, the archetypal narrative is heroic individualism, as exemplified by Lindberg's crossing the Atlantic. By contrast, in France, it is the triumph of reason as exemplified by Voltaire's Candide. In Russia, the glory of Mother Russia is celebrated in novels such as Tolstoy's (2008) *War and Peace* in 1869 and Sholokov's (1989) *And Quiet Flows the Don* written between 1932 and 1938. In each of these epics not only is a hero celebrated, but also a nation, a larger-than-life collectivity engaged in a heroic struggle, whether in the heartland or Crimea, a struggle that makes everyone great.

In all these examples a *nation* uses a narrative to make a dramatic statement about *who are we* and what is our place in the universe. Far more than simply a dry sociological text, a dramatic narrative engages the orators and the audience in a shared project of "what is our destiny," whether "once more, unto the breach" or "the Eagle has landed" The "space race" between the United States and the Soviet Union, begun in 1957 with Sputnik, and escalated with Kennedy's May 25, 1961 promise to put a man on the moon by the end of the decade, reached a climax when Neil Armstrong radioed "the Eagle has landed" from the Sea of Tranquility on July 20, 1969. This drama, as much as any armed conflict since World War II, and spanning nearly decade, affirmed the superiority of the Free World.

Nations tell stories, including heroism and sacrifice, stories which bind the nation together. A nation that has lost its sense of destiny ceases to be a nation, whether Sparta in the ancient Mediterranean or Samaria in the Middle East. These ancient nations are scarcely visible in the contemporary world. By contrast, nations that live on, whether Rome or Israel or Mother Russia, are bound together not simply by political guile but by shared stories that answer the question of "Who are we?" (Fig. 7.4).

For Americans, without question the most iconic emblem of identity is the Statue of Liberty, which, in contrast to the brazen giant the Colossus at Rhodes, proclaims inclusion, not exclusion or subjugation. Emma Lazarus's depiction of the Statue of

Fig. 7.4 The statue of liberty in New York City

Liberty as "The New Colossus"[1] as much as any text, just like the Statue of Liberty as much as any monument, makes a statement about "who we are." (Fig. 7.4)

In sum, technology and engineered achievements, both in the modern era of nuclear weapons and moon shots, and earlier engineering and architectural achievements such as the Colossus of Rhodes and the Roman Amphitheater, are ways in which societies and empires stake out their identity, their sense of "who we are." Some of these narratives and emblems, such as the Statue of Liberty or the Roman

[1] American poet Emma Lazarus wrote "The New Colossus" in 1883 to raise money for the pedestal of the Statue of Liberty. In 1903 the sonnet was engraved on a bronze plaque and mounted to the lower level of the pedestal.

Fig. 7.5 The Colossus at Rhodes according to nineteenth century engraving

Amphitheater are inclusive, while others such as Kafka's *The Castle* or the architecture of the World Trade Center, which came crashing down on September 11, 2001, are clearly exclusive and domineering. Whether the technological society has outstretched arms or a clenched fist is a central question of this era and *who we are in it*.

References

Huizinga, J. (1938/1949). *Homo Ludens*. Routledge & Kagan Paul.
Kafka, F. (1926/1992). *The Castle*. Alfred A. Knopf.

Lasch, C. (1992). The true and only heaven: Progress and its critics. W. W. Norton & Company.
Marx, L. (1964). *The machine in the garden: Technology and the pastoral ideal in America.* Oxford University Press.
Mumford, L. (1971). *Myth of the Machine: Technics and Human Development.* Mariner Books.
Norris, F. (1901). *The octopus.* Doubleday, Page.
Pfaffenberger, B. (1992). Technological dramas. *Science, Technology, & Human Values, 17*(3), 282–312.
Sholokhov, M. (1989). *And quiet flows the Don* (Vol. 1). Vintage.
Sinclair, U. (1906). *The Jungle.* Doubleday & McClure Co.
Stanislavski, K. (1950/2002). *Stanislavsky on the art of the stage.* Faber.
Tolstoy, L. (2008). *War and peace.* Vintage.
Tolstoy, L. (2016). *Anna Karenina.* Devoted Publishing.
Verne, J. (1865). *De la terre à la lune.* Pierre-Jules Hetzel.
Verne, J. (1870). *Twenty thousand leagues under the seas.* Pierre-Jules Hetzel.
Wallerstein, I. (1988). The modern world-system as a civilization. *Thesis Eleven, 20*(1), 70–86. https://doi.org/10.1177/072551368802000105

Chapter 8
Who Are We?

Abstract This chapter examines techno-totemism and other aspects of the ways in which modern society define their identity around technologies. Techno-totemism is the definition not only of one's identity but of a society's entire cosmological outlook in terms of technological objects. Cosmology, a vision of order in the universe, is an integral part of identity, of who we are, how we connect with our neighbors and kin. Cosmology unites the social and the natural, the temporal and the eternal, into a singular verity. In America, the perfect example of totemic objects is the car, yet many other examples (guns, computers, smartphones, airplanes, boats) can be found as well. Automobiles not only define *who we are*, but the entire structure of our lives and the built environment. After establishing the importance of identity and its relationship to technology, we turn to the narratives that tell the stories about who we are, but also who we are not.

In the previous chapters, we have made the case that the most important questions for any individuals or society are "Who are we?" and "Where do we fit into the universe?" Philosophers have debated this question for thousands of years and, of course, have not arrived at any definitive answer. Descartes's *cogito ergo sum* placed human reason at the center of identity, while Plato's *Republic* placed political order. The world's great religions have their own answers, each of which relates to a cosmological understanding of the universe, whether monotheistic or polytheistic, animist, or naturalist. Cosmology, a vision of order in the universe, is an integral part of identity, of who we are, how we connect with our neighbors and kin. Cosmology unites the social and the natural, the temporal and the eternal, into a singular verity. When this collapses, as it did for the Yir Yoront people of Australia's Cape York Peninsula (Sharp, 1987), the world is turned upside down.

No less than roles and statuses, institutions confer identities, a sense of "who we are" and where we fit into the universe. As noted in a previous chapter, anthropologist Mary Douglas, in *How Institutions Think* (1986), argues that in contrast to rational choice theory, the agreements and compromises on which institutions are built are cosmological building blocks for collective identities.

A. Batteau and C. Z. Miller, *Tools, Totems, and Totalities*,
https://doi.org/10.1007/978-981-97-8708-1_8

The Importance of Identity

The question of identity is a question of "Where are we starting from?" and "What is our place in the universe?" Just as a journey of a thousand miles begins with a single step, so too any cultural or philosophical construction begin with an understanding of "Where am I at right now?" For a tribal society such as the Nuer of Sudan, a fundamental element of their identity was that they were *not-Dinka* (a neighboring tribe) and in fact many tribal societies construct their identities in connection or contrast to their neighbors, either as allies or aliens. This, in fact, may be an adequate description of tribalism, an affliction that besets even some of the most advanced societies. Although the tribalisms that have afflicted American society since its founding are based primarily on race, in recent years these have been superimposed on many political differences of party, region, gender, and ideology.

Tribalism, in fact, rears its head even in the most advanced societies in the absence or derailing of a unifying national narrative. National narratives, as elaborated in the next section, are stories that *everyone* in the nation is familiar with, and which are told in many different versions in many different settings. More pointedly, there is an ongoing tension between local attachments to family and kin and narratives: the stories we tell about ourselves. In the absence of compelling narratives and storytelling, national attachments break down and tribal attachments come to the fore.

Tribalism is the opposite of patriotism, the generous attachment to the country. America as a "nation of immigrants" has historically welcomed peoples of all continents and nationalities to its shores, often with difficulty, but always with a recognition that it was immigrants and the diversity that they brought that made America great. Nearly all the technological innovations of the past century that built the American economy were by immigrants or children of immigrants, whether the automobile (perfected by Henry Ford, the son of Belgian and Irish immigrants), nuclear power (Enrico Fermi, an Italian immigrant), or the Internet (Sergei Brin, a Russian immigrant) to name a few. It is an easy demonstration that a diverse economy is more productive, adaptable, and with greater future potential than a monoculture. Tribalism is a monoculture. The lack of diversity in tribal aggregations creates a culture that is static, immobile, and intolerant.

Narrating Identity

Every nation has a story to tell, sometimes heroic, sometimes tragic. In Spain, an important part of Spanish identity is its conquest of the New World. Prior to 1492 the Kingdoms of Aragon and Castille were separate principalities, along with other subdivisions such as Barcelona and Catalonia in the Iberian Peninsula, and there was no conception of a unified Spanish identity per se. With the marriage of Ferdinand and Isabella, the kingdoms were united, and in the same year they sent an Italian navigator, Columbus, through the Pillars of Hercules across the Atlantic to the New World. Even today there are parts of the Iberian Peninsula that are asking "Who are we?" Portugal, which had its own narrative based on ocean-based exploration, was never assimilated to Spain, and provinces such as Catalonia have a shaky relationship to the rest. George Orwell's *Homage to Catalonia*, published in 1938 during the Spanish Civil War, recognized the diversity that Catalonia contributed to Spain.

Similarly, prior to 1789, there was no conception of a French nation, or even a French language per se until an uprising, spurred by the corruption of the Bourbon monarchy, overthrew Louis XVI. From the late eighteenth century into the twentieth century, defining the French nation (or the French Republic) has been an ongoing struggle, even as Charles De Gaulle in 1958 proclaimed the "Fifth Republic." As Graham Robb relates in *The Discovery of France* (2007) building the French *nation* is an ongoing project, even into the twenty-first century. Robb *explored* France on a bicycle for four years, riding more than 14,000 miles and discovering numerous cultural and linguistic variations between Paris its banlieu, between Normandy and Provence, and everything in between.

Similar stories could be told for every nation and empire. China's self-conception as the center of harmony on earth masks imperial ambitions extending all over Asia and Africa. Most recently China, through its "Belt and Road" infrastructure program, is attempting to bring the benefits of Chinese civilization to what is sometimes called "China's Second Continent" (French 2014). China sees Africa much as the United States saw the Wild West in the nineteenth century, as an uncivilized territory that would welcome harmony. China's vision of imperial glory is not simply a naked power grab, but rather a noble mission to bring harmony to a part of the planet that has been wracked by disharmony.

Storytellers—those who construct tribes and nations—have a more lasting imprint on history than kings and emperors and armies. Empires eventually are corrupted and collapse, but the stories behind them live on, even for millennia. This is found in the novels of Tolstoy, a nineteenth century Russian author, or in the *Metamorphoses* of Ovid, a Roman poet from 2000 years ago, or Mark Twain's *Life on the Mississippi* (1883).

The question of identity is important because in industrial and post-industrial societies, reflecting the churn of technological development, identity has become unstable. All that is solid melts into air, and all that is sacred is profaned (Berman, 1988). Objects and relationships that seemed to be eternal disappear or are, at least, corrupted or disrupted. The questions of "Who are we?" and "Where are we?" become especially acute in a society has seems to have lost its way and become *polarized*. The increasing tribalism of American and many other societies is adequate testimony to this. Tribalism is the statement that "not only do I not understand you, but I also don't even want to try." A great irony of the hegemony of technological society is its increased tribalism, often promoted by "social" media.

The Silences We Keep

No less important than what we talk about is what we *don't* talk about: the subjects that are off-limits, at least in polite company. This, obviously, is highly variable by culture and class: off-limits subjects in many societies include sex, money, power, religion, and politics. Many (perhaps all) institutions can be examined not only for the conversations that they foster, but also for those that they exclude. Ordinary, unremarkable silences typically include sex, family matters, shameful episodes in the past, and family members who do not fit into one's orbit.

A notable citadel of silence in America in this regard is the Claremont Institute,[1] founded in 1979 at the outset of the neoliberal revolution and in 2015 the intellectual seedbed for the takeover of the Republican Party by a cult leader. In September 2016, they published "The Flight 93 Election," which argued, in an analogy to the events on Flight 93 on September 11, that the 2016 Presidential election was a choice between terrible alternatives, but that conceding it to Trump's opponents was a death wish. Since then, the Claremont Institute has published alt-right and far-right opinion pieces but has been silent on the hundreds of thousands of deaths from COVID-19 in America, or on America's declining influence around the world.

Neoliberalism, the political movement that became ascendant in the 1980s, following the near collapse of labor unions and the increasing mobility of workers as former industrial communities began to crumble, with workers substituting careers for communal attachments. Prefigured by Hayek's *The Road to Serfdom*, published in 1944, Hayek argued that "socialism" was antithetical to the liberty on which the modern world was founded.

Friedrich Hayek argued that socialism, and more generally any overreach in government regulation of ownership would eventually result in serfdom. Interest in *The Road to Serfdom* (Hayek, 1944) took off in the 1980s, providing a charter for neoliberalism. The glaring silence of *The Road to Serfdom*, and much of the rhetoric of neoliberalism, is that it takes almost no cognizance of technology, even as information technology from the 1980s onward was building a new road to serfdom. Similarly, neoliberalism, while promising to free human energies from government domination, in fact ushered in new forms of technological domination (Fig. 8.1).

Equally notable is what Claremont *does not* talk about. Notable among these is the contributions that immigrants have made to American society. In place of celebrating immigrants as the people who have made America great, Michael Anton in "Why do we need more people in this country, anyway?" (Anton, 2018) deplored the "ceaseless importation of Third World foreigners with no tradition or, taste for, or experience in liberty." In the coming years this xenophobia would become a hallmark of the Trump presidency. "Liberty" in this context means *not* an absence of imperial

[1] The Claremont Institute, located in Upland, CA, is a well-funded conservative think tank. John C. Eastman, a senior fellow, aided former president Donald J. Trump in attempts to overturn the 2020 election. Eastman has since been criminally indicted for his role in attempting to keep Trump in office and obstructing the certification of Joe Biden's victory in the election.

Fig. 8.1 President Donald J. Trump stands before a plaque Tuesday, June 23, 2020, commemorating the 200th mile of new border wall along the U.S.-Mexico border near Yuma, Ariz (Official White House Photo by Shealah Craighead)

domination but a freedom from social restraints and a freedom to dominate the less powerful.

Also notable is the lack of a discourse around patriotism. Patriotism, in contrast to nationalism, is a love of country, not of a tribe. Patriotism is outstretched arms, not a closed fist. Steven Smith, in *Reclaiming Patriotism in an Age of Extremes,* characterizes American patriotism as "not only a statement of who we are, but also an aspiration of what we might become. To be an American is to be continually engaged in asking what it means to be an American." (Smith, 2021, p. 5) The blood and soil of contemporary American nationalism is a betrayal of more than two centuries of America's self-definition as a nation of immigrants. Keeping a subject off-limits, such as in the current example the contributions of foreigners, is no less a marker of who we are than other identity markers, and in fact in a nation whose identity has been rooted in its dynamism, the sounds of silence can be deafening.

To restate, the silences we keep are no less definitive of who we are than the stories we tell. Identity in an age of extremes forecloses narratives of shared experiences and aspirations. Nations in the technological society need to rediscover the art of storytelling, and create narratives that embrace, not reject, the diversity that made their nations great. "Huddled masses yearning to breathe free" is a narrative, an image, that for over two hundred years has united all Americans.

Where We Started

In every *nation,* there is an origin story that explains where the nation came from, and where it fits into the larger scheme of things. One of the best known of these is the story of Romulus and Remus, twins who were abandoned on the banks of the Tiber River and were suckled by a she-wolf. Eventually Romulus killed his brother and went on to found the city (and later the Empire) of Rome. In China, the creation of order (yi) out of chaos (hùndùn), and the unity or interdependence of these elements are exemplified by yang and ying trigram (Fig. 8.2), has exemplified Chinese thought ever since. In the Judeo-Christian tradition, Adam and Eve were the founders, with their sons Cain and Abel, of humankind.

In America, the origin story is the discovery of the New World and subsequent conquest of the frontier. Of course, Europeans did not "discover" the New World. Many native Americans remind us, "we were here all along," although the fact that this landmass was named, at least in the United States, after an Italian navigator is indicative of a cosmological horizon. In Canada, the aboriginal tribes are designated "first nations," indicative of a different view of the universe.

Conceptions of the universe have radically changed over the course of written history, from tribal conceptions to astronomical to metaphysical. What is embraced in the cosmos, whether members of one's tribe (and their tribal competitors), or the stars above, or the imponderable musings of the human mind and of the Creator, varies immensely among civilizations, and can (and have been) upended by the collapse of empires or the discovery of new continents. "The World Turned Upside Down", the marching tune played by English troops after the surrender of Cornwallis

Fig. 8.2 Yin and Yang

at Yorktown, is perhaps the perfect metaphor for the crises that occur when one's cosmology is disturbed.

Other cosmological crises in history have included the collapse of empires, from the Roman to the Qing to the Austro-Hungarian Empire, the Black Death in the mid-fourteenth century to the defeat of the Japanese Empire in 1945 and the election of a cult leader in America. Nations can emerge from theses crises either with rebirth, as in the example of Japan or post-Cold War Germany, or collapse.

In America, founded on the submission of enslaved Africans and native Americans to Europeans (at least in the Southern states), the emancipation of the slaves and subsequent attainment of civil rights was not simply a political contest. It was a challenge to the identity of the "master race" which still has not been fully resolved. After the end of Reconstruction in 1877, America's first domestic terrorist organization, the Ku Klux Klan, with their distinctive uniforms, became the sentries of white/black boundaries. When Barack Obama was elected President, gun sales boomed, not because of any increased physical threat to personal safety but because of a cosmological threat to personal identity. The ongoing and increasing tribalism of American society in the past two decades is indicative not just of different political interests, but of differing conceptions of American identity, in which the world is turned upside down.

Other nations have had their own cosmological crises, most typically nations with imperial ambitions that turned to dust, as all imperial ambitions eventually do. The tragedy of empire, around the globe for several millennia, has been that the end of empire is a cosmological crisis, resulting in reawakening of the question of "Who are we?" In China, the Boxer Rebellion at the end of the nineteenth century represented a cosmological crisis, as the Middle Kingdom was overrun with foreign nationals and religions. Similarly, the collapse of the Japanese Empire in 1945 brought in a resurgent technological capability that has led the world in technological innovation. In all these cases the consequence was not simply a military defeat, but a cosmological defeat, in which an existing worldview was shattered. Russia's defeat in the Russo-Japanese War of 1904–5, shattered Russia's ambitions to become a European power and ultimately paved the way to the Russian Revolution of 1917.

We have made the case that nations go through cycles of birth, collapse, and rebirth, if they have any historical memory. Since the Industrial Revolution, empires have become notably unstable. The Roman Empire lasted nearly 500 years, from the time that the Roman republic became an empire until the invasion of the Visigoths. By contrast, the Soviet Empire lasted just a little over 70 years, and the American empire is arguably now on the brink of collapse. The Industrial Revolution and subsequent post-Industrial or Technological Revolution created new instability in all of the components of identity—roles and statuses, networks, identity objects, and even categories of identity. The multiple corporate and technological empires of the past century have proven themselves to be notably unstable, under the churn of a dynamic industrial economy and the dynamics of technological innovation. In Chap. 12, "A More Brittle World," we describe how contemporary hegemonic technology has undercut the resilience of democratic institutions.

Germany's defeat in the First World War and subsequent humiliation in the Treaty of Versailles led to a national soul-searching which was (temporarily) resolved with the rise of Adolph Hitler and the Nazi Party. After Germany's defeat and partition in World War II, the country, at least West Germany, decided that its virtues were less in military conquest and more in economic innovation, leading Germany to become a powerhouse in the global economy.

This could be explored for every past or present empire, including every nation in Europe (excepting Switzerland). Empires have origin myths which explain their foundation in a glorious past. America's origin story is about the conquest of Nature in the New World, and the subduing of those referred to as "savages" in building a civilization. Although other civilizations have stories of harmony with nature, in America the founding story is the *conquest* of nature. As we have illustrated, the origin stories of empires tell a story of who we are and what is our place in the world.

In America, the origin myth is the conquest of the frontier, beginning with the settlement of the Massachusetts Bay Colony in 1620, and continuing with the War of Independence and the conquest of the West. America has always imagined itself a frontier nation, on the border between civilization and unspoiled Nature. When John F. Kennedy proclaimed the "New Frontier" in 1960, or the Electronic Frontier Foundation formed in 1990 at the dawn of the Internet era, they were drawing on a substantial American mythology. Perhaps the most iconic American figure from the twentieth century, the cowboy, represents this mythology, and numerous cowboy (and Indian) movies replayed it throughout the century. In 2020, in the wake of a bitter election, one protest group, "Cowboys for Trump," was not simply asserting an identity as cattle-herders, but rather as iconic American heroes (Slotkin, 1992).

The Places We Have Been

Part of the formation of any identity is the locations that we have explored and exploited, whether on a single continent or in outer space. When Americans first set foot on the moon in July 20, 1969, it reaffirmed not only the superiority of the United States over the Soviet Union but also a new cosmological frontier, a fulfillment of John F. Kennedy's promise from eight years earlier to put a man on the moon by the end of the decade. Kennedy's "New Frontier" signaled not simply a campaign slogan but a changing self-conception of American identity.

Nations and empires in fact define themselves by the way they bestride the narrow world. When Cassius remarks that Julius Caesar

> … he doth bestride the narrow world
> Like a Colossus, and we petty men
> Walk under his huge legs and peep about
> To find ourselves dishonourable graves.
> Man at some time are the masters of the fate:
> The fault, dear Brutus, is not in our stars,

But in ourselves that we are underlings

he is mocking not only the first Caesar with imperial ambitions but also empires in general. Numerous empires have sought to bestride the narrow world, whether the Russian empires in the nineteenth and twentieth centuries or the Austro-Hungarian Empire of the nineteenth and early twentieth century. The Mongol invasion of Europe in the fourteenth century shaped not only Europe in the late Middle Ages but also the emerging Ming Dynasty in China. From 1368 to 1644, the Mongols shaped not only Asia but also Europe. By contrast the Aztec and Inca empires, from the thirteenth to the sixteenth centuries continue to shape self-conceptions in Latin America even to this day.

One need not look at imperial glory to understand the importance of geography for self-conception. Whether Theodore Roosevelt's *The Winning of the West* (1901), or the novels of Faulkner for defining the American South in the twentieth century, these places are an important part of American identity. Christopher Clark's *Revolutionary Spring* (Clark, 2024) described how Europe was turned upside down in the multiple tumults of 1848–1849. Similarly, if more prosaically, businesses define themselves by their geographic reach. Ford Motor Company which revolutionized manufacturing in the twentieth century was an empire, with factories in Detroit, iron mines in Minnesota, and rubber plantations in South America. Like its competitor, Ford, General Electric has a global footprint in 24 countries including Thailand, Indonesia, and Norway, with revenues greater than a billion dollars.

We make the case that "the places we have been" is a central part of answering the question "who are we?" for every social formation, and "who we are not"—what is our *other*—whether "not black," or "not civilized" or "not oriental"—is an important part of "who we are." In a society beset by tribalism, narratives of "who we are not" overshadow narratives of shared experiences. Where these places and identities fit into the map of the universe is part of our relationship with the universe.

The People We Connect

Additionally, the people, both friend and foe, with whom we have relationships are a part of our definition of identity. With kinsmen and consociates this self-definition is obvious, but with members of other tribes and nationalities and races, it is also important even if it is contentious. Part of Nuer identity in southern Sudan was *not-Dinka*, and in the American South even into the twenty-first century an important part of White identity was *not-Black*. *Not-Muslim* has become increasingly important in the current century and imagined threats of the imposition of Sharia Law.[2] The principles of Islam derived from the Quran were not simply a question of the constitutional order but of the cosmological order. Related to this manifestation

[2] *Sharia* is a form of religious law and Islamic tradition that is based the Islamic scriptures, specifically, the Quran and Hadith. In Arabic the term refers to God's immutable divine law which is in contrast to *fiqh*, interpretations by Islamic scholars.

of tribal identity was witnessed recently with the spread of the COVID-19 pandemic. In some quarters, the reaction was not simply additional public health measures but also a whipping up of anti-Asian sentiment.

The major irony here is that in an increasingly globalized world, with commerce and technology bestriding an increasingly narrow world, such threats to identity are becoming more frequent. These threats may be real or perceived; regarding identity, this is a distinction without a difference. Empires today are notably unstable, and with the way technology has connected us with people like and unlike ourselves around the world, the consequence is a more unstable sense of identity. The question of "Who are we?" becomes completely unsettled in the Internet Age in which t*ribalism*, an affliction that has beset humanity since its earliest days, is ironically on the rise. Tribalism is a division of the community into those people "like us", typically based on kinship or nationality but more recently on political ideology or party affiliation or, for that matter, possibly any other immutable characteristic. "Othering" is a statement of tribalism, a position that not only do we not have anything in common with you, but I don't *want* to find any common ground.

A core part of American (White) identity is that it is not-Black. The increased tribalism of American culture, which many have remarked on, is perhaps one of the most ironic consequences of the technological society. Issues of identity that were previously discussed in coffee-houses or around family dinner tables now reverberate all over the world.

A familiarity with human history would recognize that for most of its duration, humanity has been organized into tribes. Only with the rise of civilizations have aggregations such as nations and empires mitigated tribal instincts. Nations and empires are united by shared narratives. Novelists and poets such as Hemingway and Tolstoy have done as much to advance civilization as emperors such as Nero or Caligula.

Like novelists and poets, designers, as noted in our discussion of Scott Boylston in Chap. 6, can imagine a world of diversity, designing dwellings and public spaces that are inclusive, rather than exclusive. In history, one can locate many examples of each, both inclusivity and exclusivity. As we work toward designing beyond hegemonic technology, we can focus on a society with outstretched arms rather than clenched fists.

The Objects We Treasure

Every family, community, and nation has objects in which it has invested its identity, whether family heirlooms handed down from the ancestors, public architecture, or monuments such as Mount Rushmore. These objects are statements of "who we are" no less than the stories we tell.

Perhaps the most iconic object for a shared American identity is the Statue of Liberty, a gift from the French government in 1884, designed (in part) by the notable French architect Gustav Eifel, and replicated at numerous locations around the world

Fig. 8.3 Brandenburg Gate, Germany

including Tokyo, Paris, Taipei, and dozens of other locations.[3] As described, at the base of the Statue of Liberty is inscribed a poem by Emma Lazarus, "The New Colossus," which concludes

> From her beacon-hand
> Glows world-wide welcome; her mild eyes command
> The air-bridged harbor that twin cities frame.
> "Keep, ancient lands, your storied pomp!" cries she
> With silent lips. "Give me your tired, your poor,
> Your huddled masses yearning to breathe free,
> The wretched refuse of your teeming shore.
> Send these, the homeless, tempest-tost to me,
> I lift my lamp beside the golden door!"

Most Americans who have made substantial contributions to society in the past century are descendants of the "wretched refuse," and their dynamism contributes to the ongoing dynamism of America.

Similarly for the French, the Eifel Tower built in 1887 on the centennial of the French Revolution proclaims the legacy of French culture. Like many other nationalities, whether Brazilian or Russian, Germans have erected architectural monuments to answer the question of "Who are we?" The Brandenburg Gate in Germany, built in 1788–1791 is a symbol of German unity. The centuries-long tale of unity and division, whether unified by Martin Luther in the sixteenth century or divided in the wake of World War II is a thread of that story (Figs. 8.3 and 8.4).

---

[3] https://en.wikipedia.org/wiki/replicas_of_the_statue_of_liberty

Fig. 8.4 Blueprint of the Eiffel Tower by one of its main engineers, Maurice Koechlin (ca. 1884). Size is compared to Notre Dame, the Statue of Liberty, and the Vendôme Column. Authorization given by Koechlin Family

Parsing the Universe

One analytic strategy for understanding identity, in addition to literary narratives and personal experiences is the *decomposition* of identity into its components. Broadly surveying identity over the ages, we can identify four fundamental components:

Schemata
Roles and Statuses
Networks
Identity objects

Each of these deserves elaboration. Schemata (plural of schema) are the systems of classification of people and natural objects around us, and which we admit as members of our society. Durkheim and Mauss, in *Primitive Classification* (1963), argued that societies' mental constructions reflected their social order, not reducible to the social per se, but certainly founded in it. Every society has a scheme of classification, whether based on kinship, the occupational order, or the political order,

Roles and statuses are the stuff of sociology, and can be examined in terms of kinship, nationality, organizations, class, gender, or race. Unlike networks, roles and statuses are more or less fixed and require a rite of passage to alter. Some of the central rites of passage in America and many industrial nations include school graduations, naturalization, and criminal conviction.

More recently sociology and anthropology have discovered the importance of networks for defining identity. Manuel Castell's 1996 *Rise of the Network Society* marked a pivot from a view of society as composed of classes, occupational groups, and political parties and subdivisions into one in which ties across these boundaries were increasingly a source of dynamism.

Finally, all identities revolve around identity objects, whether totems or personal possessions. "Totemism" is an anthropological concept that gained prominence with Levi-Strauss's *Le Totémisme Aujourd'hui* (Totemism Today) published in 1962. Levi-Strauss argued that the wolf clan was to the deer clan, not in terms of biology but in terms of analogy: members of the wolf clan are to members of the deer clan as wolves are to deer. More recently David Hess's concept of "technototemism" (1995) pointed out that a technological society values fewer natural objects and more technological objects—cars, obviously, but also guns and other weapons—to make a statement of "who I am." An iconic American motorcycle, the Harley-Davidson, nicknamed "the hog," (Fig. 8.5) is not simply a transportation device but a symbol of masculinity. Just as medieval guildsmen were identified by their tools, and paraded displaying them on feast days, so too modern professions make a display of their tools, whether stethoscopes, computers, spreadsheets, or law books, as a display of "Who we are." In some parts of America, acquiring firearms is an indicator of masculinity. As noted previously, in the wake of the election of an African-American as president in 2008, gun sales boomed, not because of any increase to personal safety, but because of a very real threat to American identity.

Fig. 8.5 The Harley-Davidson Hog, an embodiment of masculinity

In the recent years, as the tribalism of America has increased, the issue of firearms and restrictions on firearms has come to the forefront of public debate. The Second Amendment to the Constitution, "**A well-regulated Militia, being necessary to the security of a free State, the right of the people to keep and bear Arms, shall not be Infringed**" has in recent years been expanded to include unregulated arms far beyond flintlock muskets of the original intent, arms such as semiautomatic rifles that were nonexistent in 1791. Much of the contemporary discussion of the "original intent" of the Second Amendment authors, as reported by Shawn Hubler in the March 19 *New York Times,* (2023) is in fact less a legalistic dispute over the meaning of a statute and more a tribal conflict over American identity.

We have shown how *cosmology*—the sense of order in the universe—is an important part of the answer to the question of "Who are we?", and different nations and tribes each construct their own apparatus of sensemaking around it. When this is threatened, as shown in several of the examples above, it leads to a breakdown, a complete inability to function in multiple roles, whether productive or political.

Bringing All Back Home

One is tempted to say that all of this is totally obvious, except for the fact that in a technological society we must continually rediscover the social, that is, the importance and nature of *connections* with other members of society. When Margaret Thatcher, prime minister of Great Britain in 1987 said "there is no such thing as society" she was echoing a common sentiment of the neoliberal era that social connections are unimportant or nonexistent—a dominant ideology at a time when class interests were riding high.

More substantially, the fate of our age has been to elevate *technique*—means to ends—over ends or purposes, thus enshrining values of efficiency over values such as such as reverence or patriotism, devotion to family or community. *Building* a community is an ongoing, collective effort. The eclipse of community in America in the twentieth century, which many have documented, is one of the singular achievements of the technological society. Technological values, by contrast, represent expediency, a short-term thinking that is contrary to the spirit of patriotism.

As we begin to discover (or, more accurately, rediscover) the importance of connections and ultimate values, in a post- hegemonic technological society we can mend, reimagine, and rebuild society. This involves not simply more organizing, but also a rethinking of who we are, reckoning with our past, and constructing a new national narrative that is focused not on productivity but rather on connectivity.

References

Anton, M. (2018). Why do we need more people in this country, anyway? *The Washington Post*.

Berman, M. (1988). *All that is solid melts into air: The experience of modernity*. Penguin.

Castells, M. (1996). *The rise of the network society*. Blackwell Publishers.

Clark, C. (2024). Revolutionary spring: Europe aflame and the fight for a new world, 1848–1849, Crown.

Douglas, M. (1986). *How institutions think*. Syracuse University Press.

Durkheim, E., & Mauss, M. (1963). *Primitive classification* (Cohen & West Ltd., London, UK, 1963) (Original work published 1903).

French, H. W. (2015). *China's second continent: How a million migrants are building a new empire in Africa*. Knopf.

Hayek, F. (1944). *The road to serfdom*. University of Chicago Press.

Hess, D. (1995). *Science and technology in a multicultural world: The cultural politics of facts and artifacts*. Columbia University Press.

Hubler, S. (2023). In the gun law fights of 2023, a need for experts on the weapons of 1791. New York Times.

Levi-Strauss, C. (1962). *Totemism*. Beacon Press.

Orwell, G. (1938). *Homage to Catalonia*. Secker and Warburg.

Rob, G. (2007). *The discovery of France. Picador.*

Roosevelt, T. (1901/2021). *The winning of the west*.

Sharp, L. (1987). Steel axes for stone age Australians. In J. P. Spradley & D. W. McCurdy (Eds.), *Conformity and conflict: Readings in cultural anthropology* (6th ed., pp. 389–411). Little, Brown and Co.

Slotkin, R. (1992). Gunfighter nation: The myth of the frontier in twentieth-century America. University of Oklahoma Press.

Smith, S. B. (2021). *Reclaiming patriotism in an age of extremes*. Yale University Press.

Twain, M. (1883). *Life on the Mississippi*. James R. Osgood & Co.

Chapter 9
The Productivity Paradox

"You can see the computer age everywhere but in the productivity statistics."
Nobel Laureate Robert Solow, 1987.

Abstract Our aim in this chapter is to examine the relationship between technology and economic values, focusing on the "productivity paradox," the undisputed fact that many "labor-saving" technologies create more work, at least for the women of the society. The "productivity paradox" results from a focus on private goods at the expense of public goods and common pool resources, which has been the focus of liberal economics for the past three centuries. By considering other types of values, including both sociability and transcendence, the value added by technology becomes more subtle.

One of the primary claims for the benefits of technology is that it improves productivity, whether in the factory, at home, or on the farm. Henry Ford's assembly line made horseless carriages affordable to the masses while doubling the pay of factory workers, while Eli Whitney's cotton gin, along with the Cyrus McCormick's reaper, powered an agricultural revolution. Information technology has relieved the work of numerous typists and copyists. Similar stories could be told for hundreds of thousands of inventions since the Industrial Revolution, making modern societies overall more prosperous than earlier societies.

This prosperity, of course, is true if one counts the wealth of nations in terms of *private goods*. The focus of economics, a science that scarcely existed prior to the Industrial Revolution, is on private goods, as measured by services and industrial and agricultural output. Other sorts of goods, whether public goods such as safe streets, club goods such as craft skills, or common pool resources such as an unpolluted atmosphere, do not figure into the statistics of "productivity." The institutional context of goods such as safe streets and common pool resources, as well as club goods such as shared identities, is lost in a society that focuses exclusively on private goods.

A. Batteau and C. Z. Miller, *Tools, Totems, and Totalities*,
https://doi.org/10.1007/978-981-97-8708-1_9

It is also true if one focuses on costs and benefits that are booked, that is, recorded in *official* records, whether the books of a corporation or a family budget. Unrecorded or *externalized* costs[1] and benefits, whether in terms of "wasted" time around the water cooler or commuting or socializing, or "idle" musings about the next product breakthrough, do not figure into the calculus of economic benefits, except to be noted as "externalities," an accounting adjustment for unrecorded items. Working from home, a violation of the factory regime of the modern office, has arguably made white collar workers more productive, even if their visual supervision has escaped the office manager.

Externalities in the information age include not only the entertaining emojis that add a bit of joy that might enhance productivity, but also the wasted time waiting for networks to get up and running or repaired after a disruption. They also include the efforts to upgrade software to faster, more efficient and most entertaining versions, in theory to improve its appeal over competitors, upgrades that add little or nothing to the productivity of the enterprise. Overall, these externalities *probably* have limited the productivity of office software, a fact more experienced by the grunts and profanity emitted from their cubicles than by executives in the C-suite. In other words, the computer has shifted the burdens of productivity to frontline workers rather than easing it.

It is also worth noting that "the books," the authoritative reckoning of an organization's assets and liabilities, have a totemic status within the organizational world, defining less the actual assets and liabilities and more the boundaries between organizations (Batteau and Psenka, 2012). Any manager knows how to "cook the books," an accounting legerdemain that can make things look better, at least in the short term. "Cooking the books" in terms of confusing identities institutionally enforced (by the Internal Revenue Service) is not possible.

The technological society has altered the boundaries of productive and unproductive work. Time in a classroom is not booked as "productive," although clearly its investment in the future, certainly for the student, and probably for the entire society. Similarly, time spent in the repair shop is not counted as "productive," but it is essential for any engineered device to remain productive. The entire developmental cycle of technologies, as described by Rogers (1995), shifts the boundaries of productive and unproductive far beyond the parameters of factory production.

[1] Externalized costs are generated by producers but carried by society as a whole. Examples include air and water pollution, and more generally, climate change. Once largely overlooked, there is increasing awareness of the actual costs of production to human health, the environment, and loss of habitat for other species.

So, too, is the whole developmental cycle of technologies, a cycle that became more pronounced in the computer age. The initial stages of any technology (research and development) are *unproductive*, yet a critical stage for the development and deployment of any technology. Similarly, the later stages of the technological lifecycle beyond maturity to obsolescence and retirement are not counted in the expectations of productivity. An accounting of the entire lifecycle of any technology would yield a far less rosy picture of predicted productivity.

All technologies go through a lifecycle (as described below), starting with research and development, through ascent, through maturity and then finally through decline, as the technology becomes obsolete. The steam locomotive, for example, originating from James Watt's stationary steam engine, marked a 100-plus year transformation of rail travel, only to be succeeded by the diesel locomotive. For several decades, steam engines were *not* used for transportation, but simply for pumping water out of mines.

Further, by examining not just the core but also the peripheries of contemporary, large-scale technological systems, we discover that overall, these technologies are less about a more productive world and more about relationships between the core and the periphery. The modern world system (Wallerstein, 1988) consists of an aggregation of core nations (primarily Europe and North America, but also China), peripheral nations including most of the continent of Africa, and a semi-periphery in between. Large-scale technological systems, whether air transport or railways or power grids inevitably create imbalances between core and peripheral nations. Technological peripheries are very much a twentieth century creation as large-scale systems (notably air transport, but also the internet) have both knit the world together and created new tensions between core and periphery.

It is demonstrable that the books have a totemic status. In the previous chapter, we note that in Levi-Strauss's *Le Totémisme aujourd'hui* (1962), totems are less about ancestry or worship and more about analogical distinctions. The Eagle Clan is to the Wolf Clan as eagles are to wolves, not in terms of biology but rather in terms of analogy. So, too, the accounting records of any business define primarily the *boundaries* of the business, and not so much its entire collection of productive assets and liabilities. Assets and liabilities that do not fit into these statements are counted as externalities.

In this chapter, we focus on the externalities of a technological society, building on the previous chapters, noting that in some situations, particularly domestic situations, the externalities can loom larger than the actual booked costs and benefits. The costs of the computer age are hidden behind the books, and only are revealed in aggregate, in the entire society's overall accounting of costs and benefits.

A Paradigm Shift

The paradigm of technological benefits and productivity has been with us since the Industrial Revolution, yet now we seem to have a Copernican moment, where the paradigm might shift from the ego-centric focus on private goods to a sociocentric focus on connections and *relationality*[2]Just as in the Middle Ages, the geocentric cosmology required the addition of epicycles (Kuhn, 1962) to fine-tune the paradigm as astronomical observations improved. In the computer age, the productivity paradigm requires additional epicycles of entertainment and improved graphics and social media to account for the "waste" of the augmented computing power.

One of the major epicyclic refinements of the computer age is the increased organizational complexity. In an earlier, pre-industrial age, organizations, whether workshops or trading houses were not complex, but with the introduction of multiple layers of authority and functional specialization, complexity increased. With the computer age, organizational complexity increased, and coordination among multiple jurisdictions and institutions became a major issue and a major concern in management studies. Bradley Trainor, in *The Magical Power of Reciprocity* (Trainor, 2003), describes some of the cultural complexities in contemporary corporations. The academic specialization of "management science" is a key component of the post-industrial economy yet practically unknown before the twentieth century. The first MBA was awarded in 1908 at a time of rapid industrialization, when organizational complexity exceeded pre-industrial simplicity.

Thomas Kuhn, in *The Structure of Scientific Revolutions* (1962), observed that the Ptolemaic, or geocentric model of the universe, which survived for more than a millennium after Ptolemy developed it in the second century. It ultimately collapsed when Galileo's observations, using the telescope which he improved in 1608, replaced it with a heliocentric cosmology. This paradigm shift, in Kuhn's words, created a totally new way of looking at the universe and humanity's place in it, which as astronomical observations and instruments were improved, created a new paradigm in which the universe was far more elaborate than either Ptolemy or Galileo imagined. As newer and more refined observations are added, old paradigms collapse.

We seem to be ready for a paradigm shift in computer (hardware) and software technology, where the epicycles of entertainment and visual appeal, plus additional tasks for those workers and consumers lacking the power and social resources, eclipse actual productive value. Although a quantitative demonstration of this point would be impossible, it aligns with the experiences of many frontline workers, whether at sales terminals or call centers, whose experience with computers and digital platforms is that it takes *more* time than earlier less automated systems, a fact that is often lost in the statistics of productivity.

---

[2] According to Bourdieu (1986), social capital is one of the key elements of relationality. He asserts that subjects must be seen in context, as part of a whole. Relationality has become a major focus in design practice as systems thinking has challenged the design thinking paradigm which we discuss in Chapter 5.

What this new paradigm might look like would include an accounting for the quality and quantity of connections. On Facebook, which younger generations eschew, one can have five thousand "friends," yet in the process debase the meaning of friendship. Almost anyone will agree that a handful of trusted, intimate friends has more value that any aggregation of thousands of acquaintances ("friends") on social media, except, perhaps, for an aspiring "influencer." The contribution of the computer age to the overall quality of life and social equity requires a new paradigm, in which the assessment of the overall health of a society eclipses any tabulation of "friends" or "productivity.".

By this measure, the Computer/Information Age is wanting. The overall fractiousness and polarization of modern societies both in the U.S. and other advanced industrial nations is demonstrably exacerbated by social media (see Chap. 11 "Tweeters at the Gates"). The gatekeepers of social media have become the new emperors or warlords. To account for social cohesion or crumbling in the computer age, we need a new paradigm of productivity.

More Work for Mother

Another irony of technology is how it has shifted tasks both within the industrial and the domestic spheres, re-allocating burdens to those with less power to avoid them. Ruth Schwartz Cowan, in *More Work for Mother* (1983), observed that household technologies, whether gas stoves or washing machines, shifted onto the shoulders of housewives, tasks that had previously been performed by their husbands or servants. As standards of cleanliness improved and as domestic servants went to work in factories, the housewives, running a household and sequestered at home with the children, shouldered a greater burden.

A gendered perspective on productivity in the factory and home suggests an even starker picture of productivity. Office automation continues for the most part to confine women to poorly paid jobs in the office and factory and retailing, and computerization promotes the rationalization of all aspects of work, particularly for the less powerful.

Similarly, within the office, computers have speeded up the tabulation of data, but have make callback (help) centers *less* helpful and personal, now with machine (bot) answering the request, thus replacing a live, human person's voice and forcing customers to waste hours of idle time trying to get any problem resolved. "More work for the customer" might be considered the new paradigm of the computer age as the offloading of this unrecognized labor becomes another externality.

What we see with all these technologies is a *shifting* of the burden of productive work, most typically from the privileged roles (husbands, bosses) to the less powerful ones (housewives, servants, clerks, customers), rather than an overall diminution of labor. Of course, time spent on hold waiting to connect with an automated call center is not booked as "work" and does not figure into the statistics of "productivity." Similarly, time spent learning how to master a new gadget, such as the latest version of a software program, is *unproductive*, a fact more captured in the societal aggregate rather than in any organization's books. Overall, new technologies increase the learning curve, which affects those with less power. Many new technologies ultimately dictate to users how they must interface with a device or software program, for example, cashless systems of payment which lock out those with less access, skill, and resources. When choice is not an option, this is hegemonic technology.

Taylorism and Its Discontents

One of the major innovations in manufacturing in the twentieth century was the rise of "scientific management," the use of time-and-motion studies to optimize the man/machine interaction. Originally pioneered by Frederick Taylor, "scientific" management used clipboards and stopwatches to monitor workers' efficiency, and create standards for every action, from tightening a bolt to taking a break.

Under scientific management, the workers' actions were reduced to the smallest motions (such as picking up a wrench), standards were created for every motion, and flowcharts organized the entire production process. The workers were thus reduced to components of the machine, mechanical components who needed to know nothing other than their narrow assembly functions.

A reaction to scientific management was seen in the studies by Elton Mayo at the Hawthorne Western Electric plant outside of Chicago. Mayo tried to adjust different aspects of the work environment—lighting, temperature, ambient music, and much to the surprise of the researchers found that *any* adjustment yielded greater output. This finding, which has been called the "Hawthorne effect," (Mayo, 1979) suggests that the workers react positively to any evidence that managers are paying them attention, a social relationship *outside* the man/machine interaction. Mayo also found that one of the workers' most important motivations was their relationship with their co-workers. Current studies support that time spent socializing with co-workers, while from a microperspective counts as "unproductive," in the aggregate improves productivity.

One other reaction to scientific management was a movement of "work to rule," where production workers followed *exactly* the standards laid down, no more, no less. Through "work to rule," the "hands" could push back at management, yet not be sanctioned, because they were simply following the rules laid down by management. "Work to rule" exemplified the resources of the "hands" (workers) a fact that escaped the attention of the "brains" (bosses).

In all of this, we can see that efficiency and productivity take a back seat to authority, to the domination of the managerial class (the enforcers) over the working class, whether in the office or in the factory. One aspect of resolving the "productivity paradox" might be to recognize that *social* relationships are just as important as machine relationships, perhaps even more so as production becomes more complex. Individual workers are always more adaptable than mechanical components, unless the management instructs them not to be.

The entire discipline of "human resources," (HR) which was unknown for much of industrial history, is a perhaps belated recognition that the workers are as important as the machinery. In recent years, it has become a specialization with many firms having a vice-president for human resources, which is a testament to the criticality of the human side of the man/machine relationship. Just as we know how to fine-tune an internal combustion engine (a relatively simple device) for optimum performance, we are still discovering how to fine-tune the human factor in an increasingly complex technological environment. This fine tuning in all its myriad forms does not count in the statistics of "productivity."

Richard Buchanan's development of the "four orders of design," discussed in Chap. 5, suggests a new perspective on the computer age, in which not simply material objects but also complex systems and environments, as well as activities and organized services add up to the ensemble of productivity.

Industry's Lost Labor

Even within the modern corporation and factory, a closely supervised environment where productivity counts for everything, we can identify numerous instances of wasted resources, whether staff meetings or brand-building exercises such as exploring new marketing opportunities, that are heavily dependent on the computer and digital technologies, for example, Zoom that enables a remote global workforce. Yet these activities do not count toward the statistics of productivity. A brand, for example, is a "club good." Unlike a trademark, which is privately owned, the ownership of a brand is shared by the producers and the consumers through an interesting dynamic balance of power. There are many examples of the shared ownership of a brand. Perhaps the most startling is a Coca-Cola, a brand owned by the Coca-Cola Bottling Company. In 1985, they tried to reformulate the beverage, based on blind taste tests that indicated that consumers preferred its taste, but when it was rolled out as "New Coke," it bombed: consumers (a club) were more attached to "Classic Coke" (a club good), even if they preferred the taste of its replacement. Brands like "Coke" are club goods, not private goods.

Another example, the Harley-Davidson motorcycle referenced in the previous chapter, sometimes nicknamed the "Hog," embodies a sturdy image of masculinity, which has a corporate-sponsored group, the "Harley Owners Group," (HOG) and an annual rally in Sturgis, South Dakota. As noted above in Chap. 8, Harley owners and aficionados are not simply consumers of a transportation device, but a tribe.

"Brand equity," an accounting adjustment (epicycle) to account for the difference between "book value" (the tabulation of assets and liabilities) and "market value" (the tabulation of outstanding shares and their price). Arguably, *more* labor goes into building brand equity than into actual marketable products. *Building* a brand, as contrasted to building a tangible product, has now become a major focus of business, particularly for consumer goods, and the shifting focus of the economy in the technological society, placing less emphasis on private goods and more emphasis on public goods and club goods suggests that the productivity paradigm may well be obsolete in the technological society.

Creating New Technologies

A major issue in the productivity of technology is the introduction, ascent, and ultimately the decline and deterioration of new technologies. All technologies go through a lifecycle, in the early stages in which they are *not* productive, but in their maturity they are most productive.

In the earliest stages of the innovation process, a technology is *not* productive and may in fact swallow up billions of dollars before it reaches maturity. New innovations are constructed from antecedents which they may eventually replace in a process of creative destruction. Numerous examples of this would include anything from telephones to printed circuit boards, which after a few years of fumbling around would reveal benefits of miniaturization. The vacuum tube, invented in the late nineteenth century by Edison and Marconi, an indispensable component of any electronic device, was replaced in mid-century by the transistor, which eventually evolved to the printed circuit board, prompting Gordon Moore to formulate "Moore's Law," a techno-economic model which stated that the processing power of transistors doubled every 18 months. In the 50 or so years since he first propounded this, with current advances in microchips, it has held up quite well (Shalf, 2020) (Fig. 9.1).

One variation on the technological lifecycle is presented by Everett Rogers, in *The Diffusion of Innovations* (1995), where he argued that any technology goes through a bell curve of early adopters, majority adopters, and laggards. Different groups of users and different technologies go through different cycles of adoption. Farmers

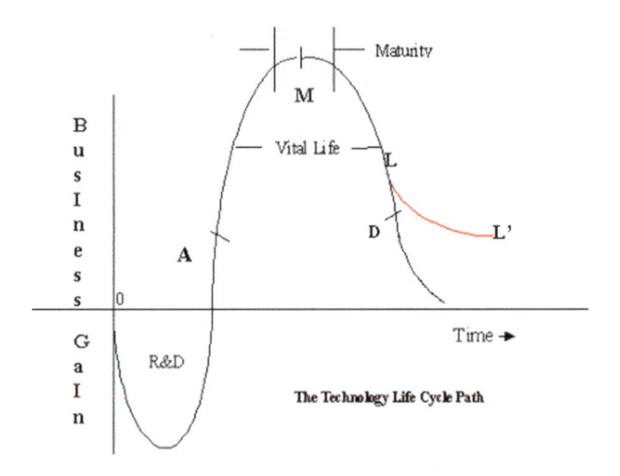

Fig. 9.1 Technology life cycle (Rogers, 1995)[3]

[3] Rogers model is still used by management consultants. All technologies exist in one of four phases:

The R&D Phase: First there are the sunk costs of identifying a problem, researching options, and developing a solution (the R&D phase). The ascent phase: Once a solution is implemented, the technology delivers a certain amount of gain. The maturity phase: Eventually, those gains level off. The decline phase: Finally, those gains are gradually exhausted as newer, more efficient technologies or processes render the solution obsolete. https://dprism.com/insights/what-istechnology-lifecycle-management/.

may be early adopters of tractors but late adopters of genetically modified crops. Truck drivers may be early adopters of more fuel-efficient engines, but late adopters of GPS-enabled navigation systems.

An important part of the technological life cycle that is often overlooked in these models is the imposition of regulation. As new technologies, whether internal combustion engines or automobiles or social media proliferate, governments scramble to keep up. Regulation, which is most typically viewed as an inconvenience by the inventors, is in fact a critical component of the social (as contrasted to the local) benefits of any technology. Whether requirements for seat-belt usage in automobiles (which demonstrably have saved many lives) or the regulation of privacy issues on social media (currently a pressing issue as social media often becomes *anti*-social media through algorithms that sow discord and invade privacy), the overall social (as contrasted to local) benefits of new technologies is a function of *society's* acceptance. Today, a pressing issue is the regulation of artificial intelligence (AI), with considerable fumbling (and political controversy) over the benefits and dangers of AI.

Given this, an obvious resolution to the productivity paradox is that the computer (hardware) and software compress the curve, or more accurately, given Moore's Law, absorb more and more innovations creating new curves that *eventually* reach maturity, but with the onrush of new technologies, with no efforts to sort out their social benefits and costs, the average user is left with the dilemma of "is this new gizmo worth the cost and learning curve?" The more typical solution to this is to ask, "What do my friends and co-workers tell me?" which brings us back to social networks.

To summarize, the technology life cycle, a well-established phenomenon spanning *all* technologies, suggests to us a new paradigm, in which social connections and relationships count for just as much, and at times more, than the mechanical or thermodynamic or cybernetic inputs and benefits of the new device. These social connections and relationships range from acquaintances made at educational institutions to memberships in craft guilds and their contemporary successors of professional associations and networks. Richard Buchanan's "four orders of design," discussed in Chap. 6, suggests a new paradigm for "productivity."

Accounting for Social Values Beyond Private Property

As the world becomes more complex and tightly coupled, "productivity" fades as a central value, possibly eclipsed by equity or resilience. Although the corporate and business world still focus on the production (and productivity) of *private* goods, with growing inequality increasingly there is a recognition in the general public that *public* goods, whether safe streets or an informed electorate are perhaps more important to the strength and overall satisfaction of the commonwealth and ultimately the individual. The ongoing tension between social equity and class domination is one of the central dramas of society in the age of hegemonic technology society.

As Elinor Ostrom described in *Governing the Commons* (1990) as well as *Rules, Games, and Common Pool Resources* (Ostrom et al., 1994) and other works, private goods are but one small corner of the overall wealth of nations, and other sorts of goods—public goods, toll goods, club goods, and common pool resources—figure in just as much. If we tabulate the different types of goods valued in human society, we find that private goods are only part overall wealth (Table 9.1).

In fact, one might infer from this tabulation that private goods are less commanding than other types of goods for the wealth of nations. Further, we find that industrial output, the aggregation that figures into productivity statistics and Gross Domestic Product, is but one of several *goods* that a society produces, most typically in the wake of the Industrial Revolution (Table 9.2).

"Productivity," a virtue associated with industrial production, came into public discussion only with the Industrial Revolution, and as Solow notes, may be fading in the computer age, post-industrial, or technological society. In the *Post-industrial Society* (1973), as Daniel Bell notes, the fact of difficult-to-quantify theoretical knowledge counts for more than material output, and the fact that something is difficult to count does *not* mean that it counts for nothing in the post-industrial society. Quite the contrary, as a matter of fact, some of the most important assets of the

Table 9.1 Goods and productivity

Type of goods and examples	How produced	"Measure of productivity"
Private goods—domestic furniture, private libraries, home pantries	Industry, human labor	Output per unit of input
Public goods—safe streets, regulation	Taxes, collective effort (e.g., safe streets)	Social accounting
Toll and club goods—craft skills, parking spaces	Organization	Collective effort, organization
Common pool resources—clean air, power grids, public libraries	Natural resources produced by nature	Cosmological

Table 9.2 Types of "Productivity"

Type of society	Productive output	How measured	Critical input(s)
Agricultural	Corn, wheat, cattle, fruits, etc	Pounds and bushels	Labor, acreage, fertilizer
Industrial	Finished goods, consumer and industrial	Productive units (e.g., cars, pounds of steel, etc.)	Labor, machinery, management
Technological	Inventions and new technologies, theoretical knowledge	Patents, trademarks, intellectual property	Education, ideas, diversity

post-industrial society, whether new understandings of nature and of society, or new social arrangements *never* figure into the statistics of productivity.

Perhaps the greatest (unplanned for) asset of the technological society is an enlarged view of the human community, an understanding that a diverse and polyglot community has far more benefits than a tribal community. Social diversity, heterogeneity, is more productive than a monoculture or homophily, even if diversity is not tabulated in the statistics of "productivity."

Accounting for Values Beyond the Core

An additional epicycle to the productivity paradigm is the creation of technological peripheries extending beyond the core. For most of history tools were local affairs, adapted to local conditions. In the pre-industrial world, architecture was based on local materials rather than global standards. Beginning in the nineteenth century, a new class of technologies, the large-scale technological system (Gras 1997) that spanned multiple ecosystems and often multiple continents, came to the fore. As noted above, the first of these on the North American continent was the Transcontinental Railway, which after the Civil War was given blocks of land to promote the settlement of the interior. Large-scale systems also included electricity, beginning with the Pearl Street Station in New York City, which went online in 1882. This system delivered electric power all over the city, and eventually to continental systems (created by Roosevelt's New Deal), with the Rural Electric power system bringing electric power to farms and homesteads which had before 1936 been powered by coal–oil lamps. This expansion unified entire continents. In the twentieth century, air transport and electronic communications (radio and then television) tied together continents and ultimately the entire world.

These technologies inevitably create cores and peripheries, much like Immanuel Wallerstein described for political systems in *The Modern World System* (1988). Wallerstein argued that the empires of old, whether Roman of Austro-Hungarian, were with modern technology being eclipsed by a system of core and peripheral regions, such as Europe and Africa, with a semi-periphery, (for example Egypt or Mexico) in between. Semi-peripheral nations and regions, which may have been part of the core in earlier times, are those who have slipped due to wars or predatory elites toward the periphery; they also include semi-peripheral nations such as Taiwan and India that through discipline and education have advanced upward in the world system.

Large-scale technologies, owing to the mismatch between rates of innovation (at the core) and rates of diffusion, create their own peripheries, in which systems that are optimized for performance at the core perform and sub-optimally on the periphery. This is illustrated, for example, by air transport, which is the safest form of transportation (in terms of fatalities per passenger-mile) available at core regions of Europe and North America, but the further one gets from these core countries the more hazardous it becomes.

What values count toward "productivity" is also heavily dependent on a society's sense of "who we are," which we discussed in Chap. 8. For agriculturalists, this is straightforward; producing crops with manual tools, agricultural output is "productive," even if research on improved crop yields or new forms of fertilizer or education in farming techniques is *not* productive. The focus of "productivity" on the here-and-now goes contrary to the entire course of technology since the Industrial Revolution, where long-range benefits such as building cities or power grids outweighed annual productivity statistics by several orders of magnitude. Indeed, an emphasis on the here-and-now makes far greater sense within the factory or workshop or farm than it does in a metropolis.

As technologies have continued to fracture the world creating fissures both within and between nations, as we will discuss in Chap. 12, "A More Brittle World," we need a new paradigm to evaluate both technologies' assets and liabilities to better understand where our world is going. We need a new frame of reference that might be like the Galilean frame of reference that enlarged the geocentric cosmology. This new paradigm would include a stronger sense of who we are, welcoming diversity and a polyglot culture rather than the contemporary class- and nationality-based paradigms. It would broaden our vision to accommodate the fact that climate change and global poverty far outweigh any benefits of material abundance and that the "productivity" of the computer-digital age could be harnessed toward the well-being of the entire planet, and not simply the material abundance of a single class.

The "computer age" introduced such a substantial paradigm shift to the industrial economy that a new paradigm is needed for understanding productivity in the post-industrial economy. The new paradigm would account for the primacy of theoretical knowledge as a foundation for deep thought, just as Galileo accounted for Earth's gravitation. This was recognized by Daniel Bell in 1973 where he argued that the service economy had eclipsed the manufacturing economy as the primary driver of productivity and that the professional and technical class had overtaken the captains of industry as the primary driver of the economy. This new paradigm would take cognizance of the *fact* that knowledge and connections and equity are far more important to a society's overall health and wealth than the accumulation of material goods. Imagining how we might design such a society based on equity, inclusion, and social justice is the subject of our conclusion. With an emphasis on participatory design and the four orders of design described by Buchanan (1992) embedding these characteristics will count for far more toward the overall benefit of society than the engineering of machinery.

The society that we see evolving beyond one dominated by hegemonic technology will reorient technology to embrace personal relationships, rather than screen relationships. Face-to-face conversation will surpass virtual life on the screen, and technology will be put in its proper place, not as the master but the servant of humanity. This will require a paradigmatic shift that rolls out to regulatory adjustments, just as the industrialization of food production as well as other goods and services led to such reforms as the Food and Drug Administration and the Fair Labor Standards Act. This is already contested territory. What these regulatory reforms might entail, including the harvesting of attention ("eyeballs") and personal information, will be one of the central dramas inaugurating a society in which the conditions and forces that engender hegemonic technology become things of the past.

References

Batteau, A. W., & Psenka, C. E. (2012). Horizons of business anthropology in a world of flexible accumulation. *Journal of Business Anthropology, 1*(1), 72–90.

Bell, D. (1973). *The coming of post-industrial society: A venture in social forecasting*. Basic Books.

Bourdieu, P. (1986). The forms of capital. In Richardson, J. G. (Ed.), *Handbook of theory and research for the sociology of education* (pp. 241–258). Greenwood.

Buchanan, R. (1992). Wicked problems in design thinking. *Design Issues, 8*(2), 5–21. https://doi.org/10.2307/1511637

Cowan, R. S. (1983). *More work for mother: The ironies of household technology from the open hearth to the microwave*. Basic Books.

Kuhn, T. (1962). *The structure of scientific revolutions*. University of Chicago Press.

Levi-Strauss, C. (1962). *Totemism*. Beacon Press.

Mayo, E. (1979). *The Social Problems of an Industrial Civilization*. London: Routledge and Kegan Paul. (Original work published 1945)

Ostrom, E. (1990). *Governing the commons: The evolution of institutions for collective action (Political Economy of Institutions and Decisions)* (1st ed.). Cambridge University Press.

Ostrom, E. (1994). *Rules, games, and common pool resources*. University of Michigan Press.

Shalf, J. (2020). The future of computing beyond Moore's Law. *Philosophical Transactions of the Royal Society A*.

Rogers, E. M. (1995). *Diffusion of innovations (4th ed.)*. Free Press.

Rogers, E. M. (2003). *Diffusion of innovations* (5th ed.). Free Press.

Trainor, B. J. (2003). The magical power of reciprocity: Or, towards an understanding of cultural integration across the division of labor in a cross-functional automotive design organization via a theory of processual culture (Publication No. 3308) [Doctoral dissertation, Wayne State University]. Wayne State University Dissertations. https://digitalcommons.wayne.edu/oa_dissertations/3308

Wallerstein, I. (1988). The modern world-system as a civilization. *Thesis Eleven, 20*(1), 70–86. https://doi.org/10.1177/072551368802000105

Chapter 10
Technology and Citizenship

Abstract In this chapter, we trace the evolution (or, perhaps, devolution) of citizenship in society under the deployment of increasingly sophisticated technology, examining how the material abundance offered by the Industrial Revolution made *greater* demands on the population to fulfill their obligations as consumers and eventually "users," a concept that reached an inflection point at the beginning of the cybernetic revolution in the 1960s, to sustain the industrial society. The interaction of the pragmatic order, now dictated by advanced technologies, with the civil order as codified on the statute books, is our primary theme. With the coming of post-industrial society, as traced by Daniel Bell in 1973, "human capital" eclipsed finance capital as the engine of prosperity, and "intellectual technology" came to the fore as the guiding force. Today, in the society dominated by technology, attention and "eyeballs" are the new circulation, and the algorithms of attention, whether on Facebook (Meta), Twitter (X), Google (Alphabet), or Amazon, are now in the saddle. *Social capital*—the investments people have in their relationships, both actual and artificial—is now a major driver of the economy.

One of the great accomplishments of the modern era is the nearly universal extension of citizenship to all classes, creeds, genders, and races. Although the concept of citizenship dates back at least to Roman times, in fact the inhabitants of medieval and early monarchies, whether the Holy Roman Empire or the British Empire were not citizens but *subjects*, with rights and responsibilities subject to the monarch's dictates. Although in the U.S. even into the second half of the Twentieth Century African Americans were de facto "second-class citizens," with voting and other rights severely constrained, sometimes by law, sometimes by custom, sometimes by intimidation, the *ideal* of citizenship was upheld for all. This tension between the *de jure* rules of citizenship (rights and responsibilities) and the *de facto* circumstances of citizens whose movements and attention are monitored in a technological society remains to be examined.

A. Batteau and C. Z. Miller, *Tools, Totems, and Totalities*,
https://doi.org/10.1007/978-981-97-8708-1_10

Citizenship is quite obviously an institutional construction, and technological threats to citizenship, as we describe in this chapter, are a consequence of what Goldin (2008) describe as "the race between education and technology." The authors provide an historical analysis of the co-evolution of educational attainment and the wage structure in the United States through the twentieth century. They argue the American educational system is what made America the richest nation in the world. Less elite than that of most European nations, by 1900 the U.S. had begun to educate the population at both the secondary and the primary school levels. This was considered a great success. However, due to a complex set of factors, this success has been disrupted by a kind of race between education, technological change, equality, and inequality.

Through the ages, concepts of citizenship have evolved, reflecting changing conceptions of the meaning of citizenship. When Cicero could proclaim "*Civis Romanus sum*" (I am a Roman citizen), he was proudly asserting an important identity. In the present technological society, the concept of citizenship has been altered, as roles of "consumers" and "users" and "spectators" eclipsed those of "citizens," not so much as a matter of law as a matter of custom, with many individuals expending far more time and effort in consumption and web surfing than in civic participation, and new corporate empires such as Google, Facebook, and Twitter (or as it is now name X) are eclipsing more traditional empires such as the U.K. or the American Empire. The imperial pretensions of major American-based global technology firms such as Amazon and Facebook, which is documented here, are discussed below. Surveillance technologies, including (but not limited to) facial recognition and tracking devices, have re-imagined the prison, only now not so much a matter of cells and stockades as cameras, cell phones, and clickbait. Similarly, attention technologies, the harvesting of attention, (Wu, 2016) are very much a twentieth century development that has been perfected in the twenty-first century, with platforms such as Facebook and Twitter (X) manipulating loyalty and attention for both political and commercial purposes. This increasing tension between the legal order of the state and the practicalities of everyday interaction on the web is the subject of this chapter. Although there are multiple foundations of this trend, including the decline of traditional institutions such as churches and civic associations, the availability of "social media" gives the trend a force and vigor that it would otherwise lack.

"Users" are the new serfs of the technological society, and Daniel Bell's, 1947 article, "Adjusting Men to Machines" argued, the imperatives of the machine age foreshadowed the need to adapt to the new technology. Bell observed that in factories, as contrasted to traditional workshops or domestic production, opportunities for face-to-face human interaction were decreased, resulting in greater dissatisfaction with the workplace (Bell, 1947). In the era of technological hegemony, we need to examine "adjusting eyeballs to the Internet" and await a Hawthorne study[1] showing the feedback loops among users and groups of users and platforms, and how these both build communities and sow divisions. Today, attention specialists aided by technology and human influencers have perfected a business model that harvests attention from literally billions of users on social media, aggregates their demographics, and sells the results to advertisers. With eyeballs fixed on small screens, face-to-face interaction is decreased, resulting in *less* personal connection. Facebook, Amazon, and numerous other corporate Internet-based empires fine-tune clickbait to capture the attention of billions of users. The erosion of civic participation this produces has been well documented.

Part of this adjustment is a changing relationship between voters and their political leaders. In earlier times, voters identified with *parties*, local communities, and workplaces. Party-defined primary elections and community-based civic organizations selected candidates for general elections. Today, there is a tribal mentality or perhaps mob psychology in party affiliation that is abetted by social media, in which algorithms determine what comes to one's attention on the screen and coalitions are created more around memes than around deep personal commitments.

In this chapter, we examine four technologies that fundamentally alter the character of citizenship: social media, surveillance systems, attention technologies, and border walls. Although civilizations for more than 2000 years have been creating structures and systems to engage and monitor citizens and restrict their movements, in the modern world these systems acquire greater force, at times enabling totalitarian regimes to extend their powers. Conversely, grass roots initiatives, whether face-to-face or on social media, have enabled citizens to connect and mobilize, counterbalancing the augmented powers of corporate entities and the state. The interplay of these new technologies, from above and from below, with the legal understandings of citizenship, redefines the meaning of citizenship and the quality of participation in the new public square.

[1] As noted previously, Mayo and the researchers who conducted the Hawthorne studies concluded that people's work performance is dependent on social issues and job satisfaction. Tangible motivators such as monetary incentives and good working conditions are generally less important in improving employee productivity than intangible motivators such as meeting individuals' desire for agency, to belong to a group and be included in decision-making and work.

Citizenship Through the Ages

The first articulation of a concept of citizenship was by the Spartans and the inhabitants of Laconia in approximately the seventh century BCE, who created a new political order that placed emphasis on self-discipline rather than political coercion. In English, the adjective "laconic" refers back to this ancient virtue of citizenship. Over the years, citizenship has waxed and waned; serfs in medieval Europe were not considered citizens but rather *subjects*; it was only with the modern revolution that the ideal of universal citizenship for all races and genders caught on, and even well into the twentieth century, this was contentious.

Classically, the built environment has provided the infrastructure of citizenship, first in terms of roads and public squares, and later in terms of postal services and printing. With the advent of the Gutenberg press and print capitalism, as traced by Benedict Anderson in *Imagined Communities* (1983), citizens residing hundreds of miles apart could *imagine* themselves as part of the same civic conversation. As media expanded beyond newspapers to include broadcast media (radio and television) and later the Internet and social media, the *imagination* of civic participation exploded. We use the term "exploded" to connote both rapid growth and also disintegration, as social media have expanded (and exploited) the tribalism of contemporary politics, a theme that is developed in section on social media below.

With the advent of broadcast media in the twentieth century, the spaces of citizenship expanded. First with radio and then television, and now with the Internet (which enables new forms of broadcasting), the conversations that defined civic participation within shared spaces among fellow citizens have evolved into clickbait and reactions to emojis. When thousands of citizens stormed the capitol on January 6, it was not in response to thoughtful discussions in coffee houses so much as it was an emotional reaction to on-line clickbait produced by algorithms calibrated to arouse emotional reaction. Emotional arousal, as many have noted, is a feature, not a bug in social media: "Keep 'em clicking" is the mantra of social media, where attention is harvested. Facebook fine-tunes its clickbait for maximum emotional arousal, irrespective of the off-line consequences. Algorithms calculate which images are likely to produce the (mostly negative) arousal. Frances Haugen, a former Facebook employee, testified before a Congressional Committee and the Securities and Exchange Commission in 2021 that it was not managerial inattention but rather deliberate corporate policy to sow negative arousal.

"Clickbait"—the images and memes that entice users (citizens?) to click on a link and follow a conversational thread—have taken off in the past fifteen years as the web has been fine-tuned to hold users' attention. Negative arousal seems to generate more mouse-clicks than positive images, the consequence being that the key to a conversation "going viral" is to go vile, stoking and arousing negative reactions. The divisiveness and tribalism which many have noted of contemporary society, of post-technological society, not just in America but around the world, is a consequence. The overall result of this is that Facebook, along with other online platforms, is the new agora, the new gathering spaces for "informed" discussions of civic affairs, only now with a *poverty* rather than a wealth of conversational resources (facial expression, tone of voice, body language, physical presence) which thus cheapens the conversation. In place of informed discussion among fellow citizens in coffee houses, we now have acrimony bouncing around the world.

Other aspects of the infrastructure of citizenship are now typically on-line: The postal service has largely been supplanted by email; similarly, shopping and the shopping mall, in the mid-twentieth century an occasion for interaction with one's fellow citizens, have now largely been replaced by Amazon, the "everything store." Amazon, which started out as a bookseller, scaled up to sell practically any consumer goods (excluding automobiles and real estate). Amazon mastered the technology of distribution much as Henry Ford a century earlier mastered the technology of production.

The net result is that the civic technology—the map, the census, the monuments—that defined civic participation and vision throughout the centuries has now gone mostly on-line. The quality of civic participation, the expectations that fellow citizens have of each other in their civic lives, becomes polluted by angry, tribal antagonisms.

An important part of citizenship through the ages has been what we might call "civic technology"—the instruments and systems that regulate and facilitate civic participation. These include not only roads and settlements, but more subtly such instruments as the census and the map. Although civilizations have been counting heads since time immemorial, it is only in the modern era that data such as ethnicity and citizenship have been included in census tabulations. Likewise, maps—two-dimensional plats of the distribution of territory and settlements—are as old as history, but it has only been since the sixteenth century that the Mercator map, a flat rendition of the Earth's (concave) surface with North represented at the top of the map (a European conceit), has become the standard. The Mercator map sorts and classified the earth's peoples in terms of geodetic distance and propinquity, sorting them into "nations" represented on the map. The complex interplay between nations (and empires) and their 2-dimensional representation is a story in itself. These technologies are today so embedded in concepts of citizenship that we scarcely give them second thought.

A consequence is that technologies such as these are defining the contours of citizenship, just as the Mercator map earlier defined the contours of the state. How *citizens* (or users, or consumers, or the mob) perceive the spaces of their civic participation affects the very quality and meaning of citizenship.

Social Media as the New Agora

The economy of connections follows the same laws as the economy of goods and services, just as Adam Smith's *The Wealth of Nations* sketched out laws of supply and demand for goods and services, and Aristotle in his *Politics* foretold the dangers of monopolies in production. Today, there are monopolies in the creation of connections, hidden behind screens and within algorithms, with the consequence that "citizens" (i.e., users) are unaware of how their attention is being manipulated, just as sharecroppers a century ago had only a dim awareness of the market forces behind their debt peonage. Sharecroppers knew their oppressors directly, but today's oppressors are less obvious.

Classically, the infrastructure of citizenship, the built environment which sustained the orderly interaction of informed citizens, consisted of several features, including public squares, roads, and urban settlements. Multiple civic activities, whether political rallies, elections, protests, or even the celebrations of sports teams, were conducted in these environments. Infrastructure also included legal apparatus for choosing elected representatives, resolving disputes, and communication apparatus (couriers, postal services). The *rhythms* of civic life revolved around these systems, which later extended to include media (first newspapers, telephones, and a few centuries later the Internet) and informal associations. All of these, obviously, have a technical foundation which, as the instrumentalities evolve, shifts. Before the beginning of railways, for example, or the building of post roads, for most citizens it was nearly impossible to imagine communication and commerce with fellow citizens in far-flung territories. Similarly, before the creation of broadcast media, whether radio or TV or now the Internet, citizens in distant localities could scarcely imagine a shared fate with their countrymen thousands of miles away.

The ancient Athenian agora, which might be considered the archetype for later gathering places, whether town squares or union halls or sports stadia, is now a virtual space: Facebook, Instagram and other social media sites are the new agora of contemporary society, and the mob psychology of Twitter (X), Reddit, or areas within the dark web, which many have commented on, are a potent political dynamic.

Social media is a phenomenon built on the hypertext transfer protocols of the Internet. Computer networks have been around since the 1950s, but when Tim Berners-Lee developed the protocols for networks to talk to each other in 1969, the social universe exploded. Early social uses of the Internet such as CompuServe specialized in point-to-point text messages, but when the first mailing list software came out in the 1990s, a new social universe was created.

This new, on-line social universe of social media has become, ironically, a place where tribalism flourishes, and its business model, the harvesting and packaging and selling of attention, has been fine-tuned to satisfy the passions of the moment. The *business model* of this new agora, a concept that was never applied to traditional public spaces, is to "keep 'em clicking," with short feedback loops resulting in tightly knit, inward-looking communities that seize on identity icons, whether objects such as puppies, kittens, guns or others such as racial groups, to stoke and seize attention. This attention is then harvested and sold to advertisers, making these attention merchants such as Facebook and Instagram some of the most valuable properties in the economy.

Surveillance, Enclosure, and Citizenship

Although slavery was abolished in the United States in 1865, for nearly a hundred years after the Civil War, Jim Crow laws constrained African Americans to second-class citizenship status, consigning them to inferior schools, neighborhoods, and public accommodations. In theory, these constraints were overturned with the Civil Rights Act of 1964 and the Voting Rights Act of 1965, but in the decades since, technological means have consigned African Americans and others to inferior, under-funded schools, employment, credit scores, and inferior housing, a phenomenon that the sociologist Ruha Benjamin (2019) calls the "New Jim Code." Often this is less through intentionality than simply from sorting by algorithm. The actors in this sorting are not conscious of the consequences of their actions. In short, technology fills in the void de facto left as the civil rights laws confined *de jure* segregation, the legal separation of groups of people based on law, to the dustbin of history.

China's "Social Credit System" (*shèhuì xìnyòng tǐxì*) is a government initiative that aggregates data from multiple sources into a centralized database containing tax records, employment history, and numerous other items of personal data including interactions with police into a central database for the purpose of so that tracking and evaluating businesses, individuals and government institutions for trustworthiness. Contemporary information technology has enabled the watchful eye of the state to be ubiquitous.

We have characterized social media and surveillance technologies as elements of hegemonic technology in the equivalent of Bell's post-industrial society. They have failed to fulfill the early promises of technology (improved productivity, improved quality of life, improved individual empowerment) but instead create connections and enclosures beyond individual agency, engrossing users (i.e., the new serfs) in service of a New Machine (Kidder, 1981). These new devices extend the powers of the state and of corporate empires *without* making measurable improvements in the quality of life or individual agency and freedom.

In Chapter One, we described the enclosure movement as the major development in European history from the seventeenth century onward in which the open fields that commoners had cultivated for generations were sequestered by lords of the manor, and the commoners were driven off into the growing industrial cities of Manchester and Amsterdam. The enclosure movement was catalytic for the budding Industrial Revolution in England and elsewhere. Three centuries later a reconstituted enclosure movement, in which public services are privatized and the agora or public square of the traditional civic commons, are replaced by the likes of Twitter (X), Facebook, and other forms of social media, "citizens" are reduced to objects of surveillance and eyeballs to be captured by clickbait. Although Facebook and other social media sites do not restrict bodily movement,[2] they do manipulate and direct the *attention* of its users, harvesting attention that can be sold to advertisers or used in other ways.

[2] Although it may be argued, they reduce the need to leave one's home for shopping, work, or meeting friends, all of which can be accomplished virtually.

These practices—from attention-harvesting to facial recognition to social media sites to virtual border walls—are technological projects, yet projects which fail to improve productivity or the flow of information or the quality of citizens' lives. Rather they tend to reinforce concentrations of power. *Citizenship*—the free interaction of citizens within a republic—is reduced to *consumer-ship*, and consumers are manipulated in ways that they are only marginally aware of. The most effective manipulation, of course, is the type where the manipulated are scarcely aware of the forces manipulating them.

In various locations around the world, citizens are pushing back, whether in the use of social media in the Arab Spring of 2011 and the Voters Not Politicians in America. Overall, the Internet and social media have created a new arena for politics, and new challenges for governing at the Federal level, which the American Constitution last amended in 1992, before the dawn of the Internet and Digital Ages.

Michel Foucault, in *Discipline and Punish* (1975), argued that the modern prison shifted the locus of punishment from the prisoners' body (torture, burning at the stake) to his soul (incarceration, perhaps forced labor). In the new surveillance state, the locus of punishment has shifted from the soul to the social life, and social isolation—the eyes of Orwell's (1949) Big Brother—is the new form of incarceration. Big Brother, in the form of social media, is likely watching.

Borders, Enclosures, and Monuments at Home and Abroad

A major technological assault on the concept of citizenship has been the growth of surveillance technology, whether facial monitoring or the growth of the carceral state. Although China's "social credit system[3]" (Hoffman, 2018), a big data system for tracking and monitoring *everyone's* movements, is an extreme example, Michel Foucault's description of the panopticon in *Discipline and Punish* (1975) points out an important current trend. The social credit system used emerging technological capabilities to extend financial credit monitoring to all aspects of life. The concept of citizenship *limited* the power of the state, yet with ample technological resources the corporate sector's power knows few limits.

In 1797, Jeremy Bentham, the founder of utilitarianism, proposed that a prison, a "panopticon," could be arranged with a single guard equipped with mirrors watching every corner of the prison (Bentham, 2017). Today, Facebook and smart phones provide the new version of the panopticon, with subjects (mostly) voluntarily or perhaps unwittingly submitting to their enclosure through their smart phones. The users' physical movements are less restricted, but their eyeballs and their desires are subtly manipulated. The extent to which citizens can or should "voluntarily" submit personal information to these platforms is a thorny question.

[3] The Australian Strategic Policy Institute International Cyber Policy Centre calls China's Social Credit System a "**Technology-enhanced authoritarian control with global consequences.**" https://www.aspi.org.au/report/social-credit.

More generally, the building of monumental border fortifications, whether Hadrian's wall in 122 CE, or the Berlin Wall in 1961, or more currently proposals for a "virtual border fence" at America's southern border, are all examples of using technology supposedly to exclude barbarian hordes, but in fact to assure the people *inside* the wall of the supreme power of the state. Although citizens are identified with a territorially demarcated state, with the growth of technological capabilities many states, particularly those with command of advanced technology, face the temptation to extend their reach into all corners of the globe. And further, "territory" is almost becoming an obsolete concept in the technological society as big data and cloud computing extend all over the globe and into space. The way Russian, Chinese, and other nefarious hackers have meddled in American elections, and how most of the world has become dependent on supply chains starting in China, are ample testimony to the obsolescence of locality in a society under technological hegemony. Arguably, the classical definition of the state, a territorially defined political unit with a monopoly on the legitimate use of force within its borders, becomes practically obsolete in an age of manipulative social media extending their reach all over the globe.

Insufficient attention has been paid to a class of technologies we refer to as *monumental* technologies, which not so subtly communicate the power of the state or a particular ideology. Although civilizations have been building monuments since antiquity, whether in the form of statues of the emperor or more recently monuments to the Lost Cause of the American South, with their visibility these create flashpoints around which different parties and factions can rally. Recent controversies over the removal of Confederate monuments in the South is testimony to this. By contrast, "virtual" border walls, surveillance systems using facial recognition technologies, send the message your location and movement are noted.

Monuments have both an aesthetic face and a political face, which in fact reinforce each other. Monuments are intentionally designed to be *imposing*—they are not tokens, but instead strong, visual, displays, typically greater than life-size. The statues to Confederate "heroes" in many American capitols are testimony to the endurance of the Lost Cause in regional mythology, and some of the most vivid controversies in America's culture wars are over the erection and placement of monuments (and the mass production of these monuments), whether to Robert E. Lee or Theodore Roosevelt. Ideally, they are aesthetically pleasing. Some of the most renowned artists have dedicated their talents to monumental creations that carry the message of reinforcing white supremacy (Fig. 10.1).

From this perspective, border fortifications should be seen not as physical barriers but rather as monumental statements, and some of the newest "monuments," notably virtual border walls (sensors either on tall towers or on drones that track movements along the border), again should be understood less from the perspective of their practicality and more from the statement that they make about the power of the state. Although physical barriers are more visually imposing, "virtual barriers" are perhaps more effective, inasmuch as you never really know who is watching; physical barriers can be evaded, but virtual barriers maybe cannot. The quantity of discussion devoted

Fig. 10.1 Robert E. Lee Monument in 2006, Lee Park, Charlottesville, VA

to a fortification on the southern border (with Mexico) suggests an obsession with these monuments.[4] (See Fig. 8.1).

Technological Hegemony: Preponderant Influence

Further, many platforms, most notably Facebook, are actively trying to assume political functions, including command of loyalty, surveillance of users, governance of public places, and media or tokens of monetary exchange. In *Facebookland* (2000) by French author Michel Bellin, there is a common currency, Libra (a blockchain medium of exchange for making payments), 3 billion citizens (only now called "users," or more recently renamed by Mark Zuckerberg "people"), a legislature (the "Oversight Board"), and borders based on the surveillance both of users and of anyone who has ever ventured onto the site. Facebookland is more like the Soviet Union, in which all economic transactions are overseen by the state, with the most critical transactions being mouse-clicks and "friends" and "likes." The only difference is that transactions are monitored not by human overseers but by algorithms, creating an edifice of automated surveillance. Like the Soviet Union, Facebook's hunger for domination has few traditional constraints. Facebook's business model rests on the harvesting of billions of users' attention, which can be sold to advertisers, making Mark Zuckerberg wealthier than any emperor in history.

Facebook is perhaps an extreme example, just as the Qing empire with its command of multiple ethnicities was an extreme example of empires around the

[4] The southern U.S. border wall has had several iterations. Although other administrations have contributed to the fence project, perhaps the best known is the Trump border wall, known as "The Wall," which was a central platform in the former president's 2016 election campaign.

world in the modern era, but in the technological world there are many other examples of technology companies overtaking the role of governments, and in the process reducing citizenship to irrelevance, an appendage of consumer-ship (or user-ship).

And further, the business model of Facebook, Twitter (X), Amazon, and others are about creating *connections*, the new currency of the technological society. Facebook, Google, Amazon, and many others harvest and trade connections, and curate their "followers" with as much care as factory managers monitored workers on assembly lines.

With the Industrial Revolution and Henry Ford's innovations in mass production, several new technologies came to the fore. As we discussed in the previous chapter, perhaps most notable of these was Frederick Taylor's "scientific management," where industrial engineers with clipboards and stopwatches monitored assembly line workers motions and set standards for nearly every step, from tightening a bolt to mounting a wheel assembly. Workers were reduced to (standardized) components of the machine, a de-skilling that many such as Harry Braverman (1998) have commented on.

As industrial production grew, the "hands" increasingly became appendages of the machines and were managed accordingly, with tightening production standards in the same spirit as tightening a screw on a machine. "Adjusting Men to Machines," a 1947 article by Daniel Bell which we discussed in Chap. 3, expresses this new sensibility of industrial production.

A notable example of corporate dominance was the rise of "company towns," in which workers, notably in mining but also in factories, lived, shopped, and socialized and even worshipped on company properties. Company towns made mockery of the medieval saying, *Stadt Luft macht frei* with debt peonage ("I owe my soul to the company store") with industrial servitude taking the place of free employment. The company town included a company store, in which miners were paid in company-issued scrip, shopped, typically at inflated prices. Until organized labor pushed back, company towns represented an apex of the industrial society in which *everything* was industrialized.

In short, just as the industrial economy rested on top of improved productivity in the agricultural economy, and the post-industrial economy rests on improved productivity in manufacturing and distribution, so, too, the Facebook era economy, which represents the hegemony of technology companies and the economy that rests on top of improved and fine-tuned productivity in the arousing and harvesting of users' attention. Eyeballs and "friends" are the new currency of the digital economy.

In 1983, Benedict Anderson, in *Imagined Communities*, described the growth of nationalism in the modern world. Although "nations" imagine themselves as extending back into the mists of a distant past, in fact it has only been in the nineteenth and twentieth centuries (roughly contemporary and coincident with the Industrial Revolution) that *nationalism*, as contrasted to tribalism or imperialism, became building blocks of the human community. For example, *nations* are assumed to be united by language, yet before 1789 most of whom we would now consider Frenchmen (residing, notably, between the Pyrenees and the English Channel, and to the west of the Rhine River), did not speak "a language," but only *un patois*. Only around l'ile de France was true "French" spoken. The grand project of building a French nation, begun in 1789, centered on L'Academie Français (founded in 1635, and suppressed in 1793 during the French Revolution) as the proper authority of the French language. Similarly, the United Kingdom, which is *united* neither by language nor religion nor ethnicity nor territorial contiguity, *imagined* itself a nation mostly in response to three technologies: the gun, the ship, and the pen (Colley, 2021). Warfare, trade, and communication *united* the U.K. far more effectively than did the throne.

Our point here is that today platforms such as Facebook are assuming more and more functions of government and other institutions: Facebook is the new agora, the meeting place for literally billions of "users" around the world, connected to their "friends" by algorithms, but potentially having nothing else in common. More accurately, what billions of "friends" on Facebook have in common is neither nationality nor any demographic, but only their Facebook membership. "Imagined friendship" in Facebookland is the new polis of a hegemonic technological society, curated and harvested by Mark Zuckerberg and his cohorts.

Citizenship and Its Threats

Classically, citizenship encompassed a set of rights and obligations to a polis, a state, or civic entity, and represented a contract, perhaps unwritten, between civic leaders and their citizens. There is always a tension, or perhaps a dynamic interplay, between the practicalities and opportunities of citizenship and the mutual rights and obligations between state and citizen. "Good citizens" do not abuse the lack of surveillance by littering streets or flouting traffic rules, even when police are not watching; good public servants maintain order and do not mis-appropriate public funds, even if they can get away with it. Newspapers, dubbed by the English philosopher Edmund Burke as the "fourth estate" for their role in representing and informing public opinion, have declined in the face of newsfeeds on the Internet. With the notable exception of a couple of newspapers of national reach (*The Washington Post* and the *New York Times*), most newspapers have either disappeared or been bought up by conglomerates such as Alden Capital. The business model of traditional newspapers, in which subscribers and advertisers supported journalism, have been rendered obsolete by the digitization of content on the Internet, and the reach of newsfeeds (as contrasted to street-corner vendors) exacerbates the polarization of America.

When one side of this balance is upended, for example by social media, then the other side must adapt, adjust, or succumb. Civic participation now (ironically) has a global reach, yet a shallowness where memes and clickbait replace face-to-face *engagement,* now not simply in one's community or country, but all over the world. A mob psychology (Le Bon, 1895) has taken over the Internet, and democratic wisdom is eclipsed by the passions of the moment.

The way numerous citizens get their information on the newsfeeds and clickbait of social media is not simply a regrettable occurrence, but rather a basic flaw of a technological society. The struggle within many states to regulate social media is indicative perhaps of the novelty of social media, or perhaps of its unique character. Should social media be regulated as a threat to consumer competition, a threat to privacy, a threat to the fundamentals of citizenship or all of the above? The "digital public sphere" that social media are constructing still requires trusted intermediate institutions. Other industries, whether automotive or television broadcast or pharmaceuticals, were initially unregulated, until the public (as represented by Congress) took note of their unique capabilities of their technologies and created regulations in the form of the Food and Drug Administration (FDA) (created as we noted previously, in 1906 in response to Upton Sinclair's *The Jungle*) or the Federal Communications Administration (created in 1934) or the National Highway Traffic Safety Administration, created in 1965 in response to Ralph Nader's *Unsafe at Any Speed* (1965), to govern emerging technologies. Ironically, the crowd psychology of social media and well-funded ranks of attorney work *against* effective government regulation. Conversely, many states around the world are using social media to undermine citizenship through the spread of mis- and disinformation. Today, the contest between generative artificial intelligence and social institutions is a frontier of technological regulation.

Efforts to regulate social media, whether in Europe or America, have foundered on an understanding of the hazards of social media. In Europe, the General Data Protection Regulation[5] (GDPR) limits the collection of personal information such as residence and employment; however, by focusing simply on issues of privacy it overlooks the *social* hazards of social media. In America, by contrast there are even fewer limits on social media, and platforms such as Facebook face no penalties for hosting not simply false but also hazardous content. The fact that each platform is left to moderate its own content and that there are no agreed-upon standards for content moderation makes these platforms less like an agora and more like the Wild West.

For an extreme example, one might look at the Peoples Republic of China, which conducts extensive surveillance of social media literally around the globe on citizens and non-citizens alike, extending the reach of the surveillance state literally to all corners of the globe. When the state becomes all-seeing, not only within its own boundaries but also around the world, then the meaning of citizenship is dramatically debased.

"Citizenship" is a legal construct, but by analyzing citizenship in *systems* terms (flows of people, information, resources, and controls, with feedback loops) will be quite instructive as to the interplay of citizenship and technology. Part of being a good citizen is staying informed regarding civic affairs, yet in the ecosystem of social media sources of misinformation and disinformation abound, and the challenges to keeping informed about public affairs is greater than ever. Algorithms drive mouse-clicks that often can reinforce existing prejudices. The capacity of Facebook to sow divisions

[5] In the European Union, the General Data Protection Regulation (GDPR) limits the power of private companies to accumulate private information. Adopted in 2016, and notably not applying to the United States, the GDPR attempts to restrain the reach of social media. The GDPR regulates both how aggregators and transmitters of personal data obtain permission from the subject before sharing the data with another party.

has been frequently commented on. Just as the technology of the printing press led to the creation of the nation-state as an "imagined community," so the technology of the Internet has led to the creation of factions and cliques and imagined "friend"-ships that are tearing the nation apart. Facebook limits its citizens to 5000 "friends" in the queue oblivious to the irony they have created. More pointedly, we might say that Facebook is illustrative of the importance of Gresham's Law, a monetary principle stating that "bad money drives out good," in Facebookland (Bellin, 2000).

We do not intend to imply any form of technological determinism here. Many dynamics, whether imperial collapse or the aggregation of territory or secessionist movements or elite corruption are implicated in undercutting the authority of the nation-state. These have existed for thousands of years, although in the past 400 years technological development has created a new tool, a new current perhaps, that adds complexity and tight coupling to these dynamics.

One need not focus on the current century to see an interplay between technology and citizenship. The steam engine and factory production hollowed out domestic life, and a century later Frederick Taylor's "scientific management" hollowed out the workplace, and in the mid-twentieth century television altered the character of civic participation. The new agora of Facebook has replaced the public square far more effectively than the Nuremberg rallies of Nazi Germany (1933 to 1938), in large part because it is more subtle.

What we see with the rise of Facebookland and other digital Internet empires is the decline and fall of the American empire. Although on the Earth's 200 million square-mile surface, there are only a limited number of physical empires, in the human imagination among Earth's 8 billion inhabitants the possibilities of imperialism are limitless.

An empire, albeit with hidden colonies and outposts on every continent (Immerwahr, 2019), including military bases, colonies such as Puerto Rico and American Samoa, and defense treaties extending around the globe, is now, like Augustinian Rome, on the verge of imploding from its own corruption and decadence. The collapse of the Roman empire in 476 CE, following corruption and infighting and subsequent Dark Ages, led to the modern revolution. The collapse of the American empire is leading first to a new Dark Ages of disorder possibly followed by a revolution that rejects preponderant influence of hegemonic technology.

Imagined Friendships in a Technological Society

Human societies are built on many different types of relationships: neighbors, family members, kinsmen, fellow citizens, friends, and colleagues. The meaning of each type is highly variable across cultures, but a constant is that society is made up of these connections, and the most important connections are enduring. Benedict Anderson, in *Imagined Communities* (1983), observed that "print capitalism," a new business model originating with the Gutenberg Press, enabled consociates living hundreds of miles apart to *imagine* themselves as belonging to a new type of community, the nation. Prior to the modern era, those residing in any given territory might conceive themselves as belonging to the same tribe, the same clan, or the same village or canton, but the idea of a nation would not arise until the future.

With the rise of social media, a new form of relationship came into being, "friendship" that could be created simply by clicking a mouse, and people began accumulating thousands of "friends." As we noted previously, Facebook limits the quantity of "friends" that can be accumulated and curated on the site to 5000, changing the meaning of friendship. Facebook thus created a new "imagined community," or more accurately an imagined empire of hundreds of thousands of "imagined communities" with attention and consumer (and political) preferences available for harvesting, albeit with only a thin gruel of evanescent commitments, and Facebookland and its progeny the Metaverse is the model empire of the technological society.

This extends to citizenship as well, less in the statutory understanding of citizenship, and more in the day-to-day reciprocity of fellow members of a community. The mob psychology of these imagined communities has been well documented, and citizenship has evolved on-line from thoughtful informed discussion and participation in courthouses and coffee houses and dinner tables into instantaneous passions of the moment. "Imagined friendship" could be a forerunner to "imagined citizenship," in which the legal, statutory construction of citizenship is eclipsed by chatbots and algorithms. The consequence of Amazon, the virtual shopping mall, is that with a click consumers can discover what their "friends" are purchasing.

In Summary

Modern technology, from social media to surveillance systems (both at the border and in the public square) to attention-driving systems of media to monumental constructions, has altered the meaning of citizenship far beyond anything that would be recognized by the Spartans or the Laconians. Although today citizens may *feel* that they are free, this simply reflects the subtlety of the forces manipulating their shopping, socializing, and even civic participation. Retail businesses have built vast empires based on their ability to manipulate and predict consumers' behavior, and political parties fine-tune their populist appeals using a modern technology (scientific surveys) to best anticipate what will arouse the masses. Political parties and factions wield these tools against each other, intentionally altering the quantity of their support, but the *quality* of participation becomes subordinated to the machinery of manipulation.

The practice of gerrymandering acquires a new tool as computers target the likeliest precincts to "crack and pack" support, diluting the strength not only of racial minorities but also any demographic identified as voting for one's opponents. By contrast, protest movements such as the Arab Spring in 2011 use the Internet to build coalitions. Citizenship is now dependent on the things that are "in the saddle," to use Emerson's words, that radically changes its meaning. The contest between citizenship from below and the unruly mob becomes sharpened in the technological society.

As civic participation becomes an appendage of what we have named "civic technology," the overall society, while optimized for the passion of the moment, becomes more brittle. Short attention spans and short feedback loops have the consequence of losing the ability to think in any manner beyond the short term, a trend that is reinforced by feedback and immediate attention of social media. In this, we find echoes of the collapse of earlier empires, whether in 476 CE with the collapse of the Roman Empire, or in 1912 with the collapse of the Qing dynasty or in 1989 with the collapse of the Soviet Empire. More recently, the collapse of multiple crypto-currency empires (as discussed in our next chapter) reveals a fixation with the dubious value of the latest technological fad. In all cases, a single-minded obsession with growth and power ultimately led to collapse, a lesson of history that the emperors of the technological society have so far failed to learn.

References

Anderson, B. (1983). *Imagined communities: Reflections on the origin and spread of nationalism.* Verso.

Balkin, J. M. (2020). How to regulate (and not regulate) social media. In *Keynote address of the Association for Computing Machinery Symposium on Computer Science and Law*, October 28, 2019. New York.

Bell, D. (1947). The study of man: Adjusting men to machines. *Commentary, 4*, 79.

Bellin, M. (2000). *Facebookland.* Kindle. French edition.

Benjamin, R. (2019). *Race after technology: Abolitionist tools for the New Jim Code.* Polity Press.

Bentham, J. (2017). *Panopticon: The inspection house.* Anodos.

Le Bon, G. (1895). *The crowd: A study of the popular mind.* International Relations and Security.

Braverman, H. (1998). *Labor and monopoly capital: The degradation of work in the twentieth century.* Monthly Review Press.

Colley, L. (2021). *The gun, the ship, and the pen: Warfare, constitutions, and the making of the modern world.* Liveright Publishing Corporation.

Foucault, M. (1975). *Discipline and punish: The birth of the prison.* Random House.

Goldin, C., & Katz, L. F. (2008). *The race between education and technology.* Belknap Press for Harvard University Press.

Immerwahr, D. (2019). *How to hide an empire: A history of the greater United States.* Farrar, Straus, and Giroux.

Kidder, T. (1981). *The soul of a new machine.* Little, Brown and Company.

Nader, R. (1965). *Unsafe at any speed: The designed-in dangers of the American Automobile.* Grossman Publishers.

Orwell, G. (1949/2003). *1984: 75th anniversary.* Berkley.

Wu, T. (2016). *The attention merchants: the epic scramble to get inside our heads.* Vintage.

Chapter 11
The Emperors' New Clothes

Abstract Empires—national, corporate, and the technological empires that are built on large-scale technological systems—are the topic of this chapter. We describe the relationship between imperial domination and decadence, the decay of ideals of humanity's salvation into profit opportunities. As large-scale technologies become dominant beginning in the nineteenth century with railways, and now in the twenty-first century embracing numerous systems including online platforms and social media, colonial domination and decadence increasingly becomes ubiquitous and taken for granted. We draw attention of the performative aspect to empires as institutions, whether in the bread and circuses of the Romans or the strutting of Elon Musk. Emperors understand at some level that they have to razzle-dazzle and delight their subjects, and staging this razzle-dazzle is part of their power. Perhaps the ultimate technological performance today is "cryptocurrency," the "mining" of virtual tokens of value and units of account through blockchain, a set of complex calculations requiring immense computing power, a technology that took off in the second decade of the twenty-first century. Cryptocurrency likewise is about razzle-dazzle more than true economic values.

Empires, large-scale aggregations of multiple states, tribes, communities, and ecosystems are as old as civilization. Empires have always coexisted uneasily with these more local systems, sometimes through tyranny, sometimes through uneasy alliances, sometimes through constitutions assuring a balance of power, sometimes hidden (Immerwahr, 2019), but always with some degree of tension. Empires colonize their outposts, placing subordinate tribes and nations under their thumb. Examples of empires through history include the Roman Empire, the Austro-Hungarian Empire (1867–1918), and the British Empire (1497–1997), and more recently the American and Soviet Russian empires. Corporate empires, as contrasted to dynastic empires, are a more recent invention, associated with the Industrial Revolution and the emergence of large-scale industry, as production shifted from craft workshops to continental industries. Examples of corporate empires from the nineteenth and early twentieth centuries include railways before they were dissolved by the Sherman Anti-trust Act, and the Power Trust until it was threatened by Roosevelt's New Deal. As industry

A. Batteau and C. Z. Miller, *Tools, Totems, and Totalities*,
https://doi.org/10.1007/978-981-97-8708-1_11

scaled up to bestride entire continents, whether through networks of railways or power grids or supply chains, local communities were increasingly at the mercy of these empires.

In the late twentieth century, a new form of corporate empire, the technological empire arose, built on large-scale technological systems, whether in transport, communication, or more recently in information technology. These are technologies—whether in rail, air transport, radio, television, or the internet—that leap over political boundaries and bring literally millions of people closer together. Today, we can conduct conversations with friends on distant continents, or plan trips to other continents, or purchase goods from anywhere in the world, all while sitting at the kitchen table with a smart phone.

There is a close relationship between imperial domination and decadence, the decay of ideals of humanity's salvation into profit opportunities. *Colonialism*—the domination and occupation of one people *and* territory by another for purposes of resource extraction, harvesting human slaves, or simply creating new markets is, of course, as old as empires, yet takes on a new aspect, not fully appreciated, in the shadow of technological empires. Whether the Industrial Revolution which tethered millions of factory workers to their jobs and their residence in company towns, or the internet which tethered billions of consumers to their smart phones, in all cases an empire is a large-scale domination of labor, consumption, attention, and loyalty, and sends a message to the colonial subjects that "you matter far less than our colonial objectives." As large-scale technologies become dominant (hegemonic) beginning in the nineteenth century with railways, and now in the twenty-first century embracing numerous systems including online platforms and social media, colonial domination and decadence increasingly becomes ubiquitous and taken for granted.

We should make it clear that with respect to these large-scale technologies, "empire" is not so much a figure of speech as it is a literal description, only now harvesting users in the form of attention and personal data and wealth, including likes/dislikes, networks, connections, and locations. Attention has replaced real estate as an object of exploitation, and the smart phone user's browsing history is a gold mine of information for profiling the user for targeted advertising. Political allegiance and loyalty are part of this picture. Empires rise and fall, yet ruling corporate and technological empires are demonstrably more brittle and failure-prone than earlier territorial empires.

We should also make clear that there is a performative aspect to empires as institutions, whether in the bread and circuses of the Romans or the strutting of Elon Musk. Emperors understand at some level that they have to razzle-dazzle and delight their subjects, and staging this razzle-dazzle is part of their power.

A Technological Empire at the Close of the Twentieth Century

Perhaps the avatar of this new hegemonic technological world, in many ways was Jack Welch, the CEO of General Electric from 1981 to 2001. Jack Welch took GE from a manufacturer of electric hardware in the mid-twentieth century into a conglomerate at the close of the century, pioneering a new form of business, large-scale financial services. In the process of transforming GE, Welch closed factories, laid off employees, gutted communities, and eventually gutted the brand. Parallel to Welch's reign at GE, and in fact reinforced by it, was the rise of "financial engineering," or "fintech," a business model unknown for most of the twentieth century yet now a dominant player in capital markets. Financial engineering is the management of assets and liabilities and human resources and cash flows with the explicit purpose of improving the bottom line and shareholder value,[1] typically at the expense of other values. A corporation provides and embraces numerous values—a brand, a culture, a community, a legacy—and optimizes this complexity for the least-common-denominator of cash flow, although difficult in a competitive market (unless the competition can be eliminated), is intellectually less challenging than embracing the entire array of corporate complexity. The movement for "shareholder value" was unknown before the neoliberal era beginning around 1980, yet now is a driving force in business. This rejection of complexity in favor of the simplicity of the bottom line in a world that is becoming increasingly complex due to globalization and technological innovation is perhaps the ultimate anti-social statement. Eventually, this simplification of the business makes it less adaptable (a subject that we treat in the next chapter), substituting financial and technological leadership for social leadership in an ever more connected yet more diverse world.

Leadership—the inspiring of followers, the maintenance of order, and the allocation of rewards and punishments, whether by a general on the battlefield or an orator in the town square or a CEO in the boardroom—is an essential part of any collective effort. Leadership embodies an embrace of the complexity, both of the organization and of its environment. Social leadership involves a comprehensive understanding and awareness of the complexity of the social milieu. However, when it is degraded to technique, at times in an effort to please multiple and diverse stakeholders, much of this complexity is lost. When the leader has a singular focus (the "bottom line"), the organization has a singular focus that results in an unsustainable business model in an increasingly complex world. The recent movement toward environmental, social, and governance (ESG) values suggests an enlargement of the corporate vision, yet corporate concern for ESG values is contested as inappropriate by neoliberal pundits

[1] Shareholder value is counterpoised to "stakeholder value," the recognition that numerous individuals and communities have an *investment* in the business, even if that investment is not monetized. Workers, communities, loyal customers, all have, to varying degrees, committed their lives, their identities, and their loyalties to the business, an investment that is discounted in the calculation of "shareholder value." As obvious as this is, it is all dissolved in the icy bath of shareholder value calculation.

who argue that shareholder value should be a firm's primary objective. This is one of the central dramas of today's technological society.

Over 20 years, Jack Welch transformed General Electric from a pillar of American manufacturing into a financial services firm, in the process enriching thousands of shareholders (and leaving himself with a generous retirement) yet hollowing out dozens of communities with closed factories and thousands of laid off workers. Perhaps the ultimate coda for Welch's tenure was the roll-out of the Boeing 737 Max which, after two spectacular crashes in 2019 and 2020, killing all 346 aboard, was taken out of production.[2] Welch transformed Boeing, a GE subsidiary, from a leading aerospace firm into a symbol of all that could go wrong in high-tech. Numerous other technological disasters, whether Three Mile Island or Chernobyl, and less dramatic disasters such as Facebook also echo the GE template which we discuss in the following section.

Roger McNamee, one of the original investors in Facebook, in his 2019 book *Zucked: Waking Up to the Facebook Catastrophe*, describes the disastrous consequences of Facebook's decision to monetize *everything*. As McNamee describes it, it was not simply a technological evolution but rather a series of discreet design decisions that created the social media platform that is now Facebook (Bissell, 2019). These decisions include the choice of colors and emojis for arousal, the algorithms which steer users to emotion-arousing content, and the decision noted in Chapter 10 to "limit" Facebook users to 5000 "friends." Although multiple social media platforms similarly manipulate users' attention, Facebook is an extreme example of *successfully* monetizing and monopolizing attention in the attention economy even though younger generations have turned their attention to other platforms. The acquisition of Instagram by Facebook was a successful strategic move to maintain relevance.

After Welch left, GE as a corporate empire collapsed into a holding company and was eventually broken up into three companies: aviation, power turbines, and medical equipment. While each of these can be a viable business, combining them (plus financial services plus consumer electronics plus aerospace) into a single conglomerate was an effort to circumvent anti-trust legislation forbidding monopolies. Conglomerates, particularly in unrelated sectors, are a late twentieth-century business model and have become the new business empires.

In all of this, we can see Jack Welch and his tenure at GE as a roadmap for other industries, particularly those that traded in twentieth-century solid goods, rather than media or information. The first wave of industrialization from the nineteenth century up to the Neoliberal Era was about tangible goods and services, not information or media. With the coming of post-industrial or technological society, these old-line industries were eclipsed by media and information industries. With the collapse of GE, we can possibly foresee the collapse of the technological society as its decadence

[2] A recent update to Boeing's ongoing story covered by the Washington Post. "Boeing's decision to plead guilty to a felony fraud charge for its role in two plane crashes that killed 346 people marks an effort to open a new chapter after a half-decade of tumult and investigations. But immediate reactions to the deal – unveiled shortly before midnight Sunday – suggests moving on won't be easy.".

https://www.washingtonpost.com/business/2024/07/08/boeing-guilty-fraud-crash/

is revealed. Part of GE's collapse was the manipulation of earnings to meet quarterly expectations (Gelles, 2022). As any manager knows, monthly or quarterly financials can be massaged by hiding liabilities, perhaps inflating the value of assets, taking advance credit for income or deferring liabilities, and otherwise making the numbers look good. As a one-off, this is not difficult, but sustained over decades eventually results in collapse, which indeed was the fate of General Electric after Welch's departure.

Following in Jack Welch's Footsteps

Others that have followed the GE template include Jeff Bezos, CEO of Amazon, Mark Zuckerberg of Facebook noted previously, Elon Musk and Twitter, and Kenneth Lay and Enron. Included on this list is Samuel Bankman-Fried and the lords of cryptocurrency. Jeff Bezos, for example, launched Amazon in his garage as an online bookseller in 1994, just as the internet was taking off. Books are a tangible and identifiable product, but after a few years Bezos discovered that his online model could be scaled to sell supposedly anything, earning Amazon the sobriquet of the "everything store." Ventures that Amazon has now entered include groceries (Whole Foods), cloud computing (Amazon Web Services), venture capital, retail distribution from many manufacturers, and many others including some that have been attempted and later abandoned. Amazon supports its distribution with hundreds of "fulfillment centers" around the world (28 in California alone), which mirror Blake's "dark satanic mills," albeit with improved lighting and air conditioning.

Over the space of 25 years, Amazon went from a bookseller to one of the world's most valuable corporations (market capitalization of nearly a trillion dollars), yet in the process gutting the vitality of central cities and their shopping districts. The ambient commons, the rich, unstructured landscape of information which characterized much of urban life in the twentieth century has been disrupted by cell phones and mouse clicks.

Perhaps the most recent prodigy of Jack Welch is Samuel Bankman-Fried, whose cryptocurrency empire FTX had a market value of several billion dollars before it crashed. Bankman-Fried was arrested in 2022 in the Bahamas, a notorious haven for money-laundering. In Bankman-Fried, we can see not only a disdain for the law but also for social niceties of life-in-person such as personal appearance (Fig. 11.1).

Most recently Amazon's venture into space travel is its subsidiary Blue Origin, which currently is focused on space tourism and sub-orbital flights. In partnership with NASA, along with other space ventures such as the Sierra Nevada Corporation, Boeing, Virgin Galatic, Red Wire Space, Blue Origin intends to promote space travel, sub-orbital and beyond, as the next colonial frontier of America.

The allure of space travel draws on science fiction, from, for example, the 1865 Jules Verne classic, "De la terre a la lune," and to the moon mission of the 1960s and 1970s. It also draws on the 500-year history of European colonialism in Africa and the Americas, in which European nations beginning with the kingdoms of Aragon and

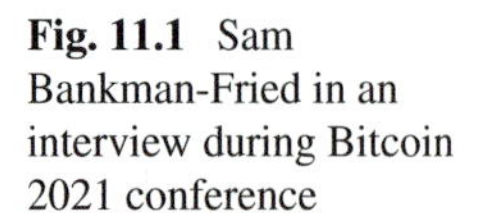

Fig. 11.1 Sam Bankman-Fried in an interview during Bitcoin 2021 conference

Castile and later Great Britain and France and still later the Netherlands and Belgium built empires beginning in the New World but extending to Africa and Asia. There were multiple reasons that the Spanish and English and French imperial adventures were successful. First, there was a wealth of natural and human resources to harvest, second, these could be economically harvested, whether as crops or slaves, and third, it fulfilled their imperial ambitions. Only the final of these apply to contemporary expeditions to Mars or the Moon. It will take sustained attention over decades or even centuries of imperial rise and collapse, changing national priorities, and notoriously short attention span of contemporary societies to harvest anything of value from Mars.

"Harvesting," of course, is as old as agriculture and civilization, yet the dream of harvesting either natural bounty or human slaves from the wilderness beyond civilized frontiers is an imperial project, whether the Romans harvesting slaves from southern Gaul or the British harvesting slaves from Africa, or Facebook harvesting the attention of billions of users for sale to advertisers. In both cases, it is the *nec plus ultra*—nothing further beyond—of imperialism, a project potentially discredited after World War II yet still alive in the techno-verse and now the emerging metaverse. Technology has enabled new forms of imperialism, either with new tools of conquest and riches to harvest, or with new robes of grandeur for the emperor. Space travel beyond the low Earth orbit of satellites is *not* a viable economic proposition but rather an imperial gamble, cloaked in the new robes of technology, yet available only to techno-titans such as Jeff Bezos and Elon Musk.

Technology, in short, has invented and is now perfecting new forms of colonialism, based less on harvesting natural resources and more on harvesting social resources including attention, shopping habits, personal networks, and personal preferences. These are then used to manipulate the addictive tendencies of the technology into further addictive behavior.

Beyond Earth's Orbit: The Metaverse

Another techno dream is the "metaverse," promoted by Mark Zuckerberg, CEO of Meta and its subsidiary Facebook which we discussed in the previous chapter. Facebook, which was launched as a student experience at Harvard in February 2004, quickly discovered that data on its users' connections and preferences—"friends" and "likes"—could be aggregated, harvested, and sold. Most recently Facebook has ventured into the "metaverse," an alternative reality where "users" (both homo sapiens and bots) can connect, communicate, engage in commerce, and form "friendships" in the "metaverse." The metaverse is a synthetic reality, requiring computer screens and special headsets to immerse oneself in its sensory experience (Fig. 11.2).

In the metaverse, one can travel to other planets, enjoy fine wine, have sex, cuddle with furry animals, and listen to an opera all while located comfortably anywhere. At least this is the dream. There are, in fact, numerous real-world obstacles to scaling up the metaverse: lack of computer processing power, training for users, potential effects on health, availability of hardware and bandwidth, and reach of the networks. These physical limitations will *always* encumber large-scale systems, whose ambitions

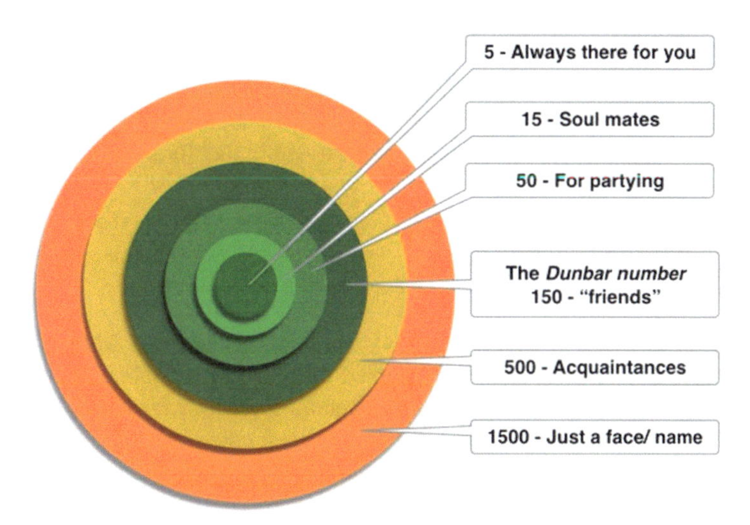

Fig. 11.2 Dunbar's number (Quora)

always exceed their physical limitations, and the dreamy-eyed aspirations of techno-titans inevitably sour. Even more important are numerous social issues with personal privacy at the top of the list. Due to the gutting of agencies that would enforce privacy protection, unless Congress finally steps up to challenge Big Tech, privacy concerns will be quickly overcome.

Facebook harvests and aggregates terabytes of user data for sale to advertisers, making it one of the most valuable companies (in terms of market capitalization) in the world. When one enters the Metaverse, the eyes of Big Brother, George Orwell's characterization of totalitarian government in his novel *1984*, are ubiquitous. Privacy is fast becoming a major issue in the technological society although many of younger generations, so called digital natives who have been raised in the digital world consider the transparency of their private lives as normal. However, at the same time the urgency around privacy is increasing as so much has already been traded away. Big Brother was an amateur, and in fact China's "social credit system" described in the previous chapter represents only the minimum of possible surveillance technologies. The Chinese social media platform TikTok, owned by Bytedance, is under investigation for sweeping up personal data on its users. The controversies surrounding TikTok would merit an entire volume: TikTok more so than any other platform (except possibly Facebook) has created an alternative *universe* in which attention, identity, and revenue are absorbed by a platform owned by a hostile state power.

TikTok and Facebook may well signal the collapse of the hegemony of the technological society. As citizens become less enchanted with invasions of their privacy and manipulations of their attention, they may turn to other sources of value, such as relationships. In the unlikely event that the Metaverse is scaled up, it will primarily create new fissures within the world between core nations and regions, which have the computing processing power and skills and privacy safeguards, and the periphery where these are lacking. In short, like so much else in the realm of large-scale systems, far from bringing the world closer together it will in fact further fracture the world.

Tweeters at the Gates

Finally, in this parade of techno-titans, we must include Elon Musk, the founder of the electric car company Tesla (named after one of the pioneers of electricity, Nikolai Tesla), and now an owner of Twitter, most recently renamed "X." Twitter, a social technology platform, presents itself as a technological frontier is arguably an anti-social technology. By limiting messages to 280 characters including emojis and CAPITAL LETTERS, Twitter substitutes drunken barroom shouting (both among real people and bots) for thoughtful political discourse. This dumbing down of democracy, a dumbing down led by one of Twitter's most prominent users before he was kicked off the platform, is a current and pressing issue. "Truth Social," the Orwellian alternative to Twitter (X) owned by a former U.S. president, is now attracting its own tribe.

Twitter, along with Facebook and now TikTok, has become the new town square, on which (subject to mild moderation) a free flow of messages among users leads to the formation of collective opinions. However, like Facebook, Twitter has an immense database of algorithms to encourage and spread those messages that are most likely to arouse emotions, more typically negative emotions. Social harmony is far more difficult to cultivate than shouting matches, and when it is drowned out amid a torrent of negative tweets, it dissolves altogether. Not unlike the Visigoths who sacked Rome in 410 CE, the tweeters are now sacking democracy, and one of the world's richest men, Elon Musk, is seeking to own the town square, but in fact may be destroying it.

Elon Musk is writing the latest chapter in the drama and theatricality of technological imperialism. In addition to his many companies including Tesla and the Boring Company (that builds tunnels for transportation systems and utilities), he cofounded a satirical monthly, *Thud*,[3] patterned off *The Onion* (another satirical publication) and several other business and technological entities. Musk who has earned acclaim for his inventiveness, typifies other technological emperors both in terms of his technological and business ventures, both successful and unsuccessful, and in his outlandish displays that flaunt both his wealth and his idiosyncratic tastes.

Thoughtful conversation requires multiple communicative channels—words, eye contact, tone of voice, body language, phrasing—that are completely unavailable on Twitter. Twitter demonstrably often substitutes thought*less* discussion for thought*ful* discourse. If the subject of the discussion is yesterday's weather or a recent sports event, this is often harmless. If the subject is a constitutional crisis, the fate of an election, or the validity of science, it has the potential to drive a dagger through the heart of civil discourse and democracy.

The Decline and Fall of Monetary Empires

Perhaps the ultimate technological performance today is "cryptocurrency," the "mining" of virtual tokens of value and units of account through blockchain, a set of complex calculations requiring immense computing power, a technology that took off in the second decade of the twenty-first century. Cryptocurrency is inherently worthless, but if the promoters can convince enough willing purchasers that it has value, it *will* have value, at least until it does not.

There is a finite supply of cryptocurrency tokens, which can be "unearthed" through complex computations and then stored on private servers. Cryptocurrency disrupts the history of money and has created a new market. Crypto has the potential to disrupt national currencies by freeing value from social and institutional restraints (as depicted in Fig. 11.3). When value is based on tangible assets, such as gold bullion as currencies were when the gold standard was in place, it is difficult to compromise,

[3] "Thud" which Musk co-founded in 2017 went defunct in 2019 after he pulled his funding.

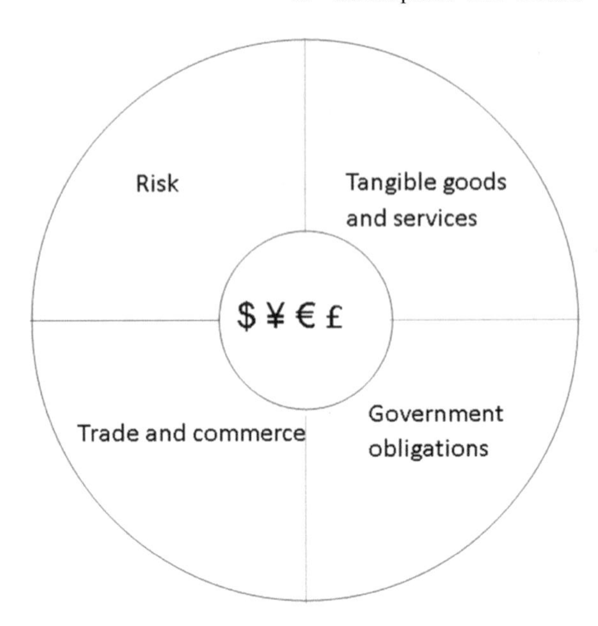

Fig. 11.3 Institutional context of money

although the market value will fluctuate; when it is based on social commitments, such as banknotes, without active social oversight it can be debased.

Money supplies *liquidity*—the free flow of value among incommensurables including tangible goods and services, trade and commerce, risk, and government obligations (both to and from citizens) (Fig. 11.3).

Each of these exists in an institutional and historical context, whether the laws (and law enforcement apparatus) protecting private property including tangible goods, or the insurance industry managing risk, or the productive firms that promote trade and commerce. Institutions are the brick and mortar of society, in which members of society have invested not only their resources but also their identities—their sense of "who we are." Whenever these are threatened, whether by wars or natural disasters or pandemics or constitutional crises or corruption, other institutions step in to heal the breach or else the entire society collapses, whether the fall of the Roman Empire in the fifth century CE or the waning of the Middle Ages in the fourteenth century or the collapse of the Bourbon monarchy in 1789. Money ($, ¥, €, £) is the medium of circulation within this orbit, commensurating the incommensurables of for example risk and government. Money *lubricates* the system; it does not *fuel* it. Money does not "make the world go 'round." Pumping more money into the system at best adds instability and more typically crashes the system entirely. This has been the story of many bubbles, whether the Dutch tulip bulb bubble in 1637 when speculators bid up the price of tulip bulbs or the stock market crash in 1929, and in 2007–2008 with uncollateralized loans. In the past 40 years, bubbles have included the dot-com crash of 2000–2002, or the sub-prime mortgage crash of 2008. In all these cases, get-rich-quick schemes, enabled by technology, simply benefit those who think they can outsmart their fellow citizens.

This is indeed the history of money for the first century of America, as banks and various governing entities took on obligations (and sometimes defaulted) with no oversight. Alexander Hamilton, America's first Secretary of the Treasury, took a decisive step in unifying the newly independent United States by creating the First Bank of the United States, and assuming the Revolutionary War debts of the thirteen former colonies. Otherwise, the thirteen colonies would have never become the *United* States. During the Civil War, the power of banking, both North and South, were challenged, yet arguably the leading reason why the North won was due to a stronger industrial economy than the slaveholding South. Only with the creation of the Federal Reserve Bank system in 1913 with the objective of stabilizing the banking system, subsequent reforms of the New Deal (notably the Glass-Steagall Act of 1933, which separated commercial from investment banking, and was repealed in 1999 at the height of the neoliberal era) was America's money stabilized. Money, in short, is always embedded in a society's institutions. When those institutions, whether the housing market or the stock market, prove themselves to be faithless, trust is broken and the value of money collapses.

Cryptocurrency is arguably becoming apparent as a Ponzi scheme, in which maintaining its value is based less on any inherent value, and more on recruiting a growing number of gullible speculators ("investors"). Examples of Ponzi schemes and speculative bubbles in the modern era include Bernie Madoff's Investment Securities which swindled hundreds of millions of dollars from investors before his arrest in 2008, and most recently StableCoin[4] which was to be tethered to the dollar but in fact has lost most of its value. Ponzi schemes are very much a business model of the technological society, enabled by technological propagation. More generally, we can observe that institutions are the guardrails of a society, amid the churn and turnover of its larger environment. In a dynamic society, risks and opportunities are always emerging, and without an institutional foundation, the society will crumble. The history of money over the past few centuries illustrates this, as "investors" tried to outsmart their fellow citizens, usually creating bubbles and crashes that made everyone poorer. Today technology has created new possibilities for creating bubbles, impoverishing the entire society. Georg Simmel, in the Philosophy of Money (1900/2011), makes clear that money is ultimately a *social* project.

Performance vs. Productivity

A common thread through all of this, and indeed the avatar of the technological world is that these cutting-edge technologies are more about performance rather than productivity. Only a small percentage of those who had "invested" in cryptocurrency can explain its workings; they simply see that its value is going up, and FOMO ("fear

[4] Senator Elizabeth Warren expressed her concerns about StableCoin in a letter to the House Financial Services Committee on April 8, 2024. https://www.warren.senate.gov/imo/media/doc/letter_to_hfsc_on_stablecoins.pdf.

of missing out") leads them to jump on the bandwagon. Not unlike the Dutch Tulip Bulb mania of the seventeenth century, a herd mentality overcomes rational thought.

Indeed, we are coming to recognize that performance and theatricality are as important social dynamics as production, distribution, and consumption. This was evident in the bread and circuses of the early Roman Empire, and the theatricality of contemporary North Korea, and now the drama of Elon Musk's (self-proclaimed as the "Chief Twit") takeover of Twitter, where he arrived the first day at the Twitter office carrying a kitchen sink (Fig. 11.4). What is novel in recent years is *technology* as theatricality, and more recently, the realization that in many instances technology is *only* about theatricality. Most theories of technology and society focus on technology's contribution to improved productivity and distribution; in the contemporary world, there needs to be a focus on hegemony of technology and its contribution to domination and subjugation.

The benefits of technology, which have pre-occupied us for more than a century, are now fading before its theatricality, at least in terms of civil society. While we recognize the benefits, we insist that they are more complex than simply increased productivity and leisure (see Chap. 9, "The Productivity Paradox"). We need a more multi-dimensional and nuanced understanding of "technology" than simply sophisticated instrumentalities.

Analyzing technology through the lens of the ritual process and theatricality can be revealing not only regarding its mechanical and thermodynamic properties but also its theatrical properties. Bryan Pfaffenberger's concept of "technological drama" (1992) can be expanded to include not only the developmental cycle of new technologies but also their place in imperial domination. Ritual processes such as counter-signification, counter-appropriation, and counter-delegation, in which the

Fig. 11.4 Chief Twit's first day at office

technologies' social role and acceptance go much further than their practical utility. In Space travel for example, to understand Blue Origin's appeal: when a retailer and son of Cuban immigrants can proclaim himself a master of the sub-orbital universe, he is stepping onto a stage to announce that "I have arrived."

Seen in this light, the hollow promises of techno-titans are but one instance of a more general corruption that has beset America. Technology can now be added to a long litany of gods that failed, including communism, liberalism, and even democracy. *The God that Failed* (Crossman, 2001), a collection of essays by disillusioned ex-communists including Andre Gide, Arthur Koestler, and others, describes how in the twentieth century Communism went from a messianic cult to a totalitarian prison. Koestler, who grew up in the twilight of the Austro-Hungarian Empire, lived through the corruption and collapse of both that empire and the subsequent aspiring empires of Nazi Germany and Soviet Russia. Koestler was aware of the dangers of a shallow understanding of technology:

> Modern man lives isolated in his artificial environment, not because the artificial is evil as such, but because of his lack of comprehension of the forces which make it work – of the principles which relate his gadgets to the forces of nature, to the universal order. It is not central heating which makes his existence 'unnatural,' but his refusal to take an interest in the principles behind it. By being entirely dependent on science, yet closing his mind to it, he leads the life of an urban barbarian. (Koestler, 1968, p. 264)

While corruption is as old as civilization and empires, the Industrial Revolution has now scaled it up and given it additional force. Indeed, a general theory of society and politics suited for today should have a comprehension of corruption and misappropriation at its core.

In all of this, we see a common thread of technology as performance. Erving Goffman's *The Presentation of Self in Everyday Life* (1959) first introduced a performative aspect into the social sciences.

> The self, then, as a performed character, is not an organic thing that has a specific location, whose fundamental fate is to be born, to mature, and to die; it is a dramatic effect arising diffusely from a scene that is presented, and the characteristic issue, the crucial concern, is whether it will be credited or discredited. (p. 252-253)

In a technological society, there are even more resources for performance. Technology, which was supposed to improve productivity, make life better and the world smaller, instead is now used to proclaim the grandeur the titans of the industry. Space travel, for example, unlike the colonization of Africa or the Americas at the beginning of the modern era, will arguably not yield natural resources, new customers or human resources, or new markets; space travel is primarily about display. The thought that space travel could yield any of this assumes that sufficient attention and investment can be sustained, not over decades but over centuries, in an era where empires are notably collapsing.

Cryptocurrency likewise is about razzle-dazzle more than true economic values. Money is deeply implicated in society's institutions and tangible assets, whether the banking system, real estate, or the royal treasury. Cryptocurrency purports to untether "money" from institutions, whereas, as previously noted above, money is the lubricant of institutions. Nobody has paid his taxes in cryptocurrency nor has any government (or corporation) settled its payroll with cryptocurrency. The stability of these institutions providing employment, income, public safety and other public goods, depends on their ability to forge compromises with other institutions and individuals, *not* on any illusory ability to dominate the social landscape. *Compromise*—the art of "let's get along together"—depends on a willingness *not* to pursue domination, but rather on a willingness to find a common good in an interpersonal relationship. Building the common good is the long story of human history, but in a technological society, ideas of the common good fade before concepts of efficiency and gratification. In the techno-verse, domination is a dream and effortless gratification is pursued by millions. True gratification is *earned*, and when it is effortless, it can just as easily fade away.

In all of these—space travel, the creation of alternative realities, cryptocurrency, or even "social" media—we see a fragmenting of the world through a common thread of technology. The common threads connecting all of these is first a promise of effortless gratification, and second a magical end to social woes. The *magic* of technology, which has enchanted millions and which we elaborate on in Chap. 13, signals the next-to-final chapter of the technological society.

Corruption and Decadence

Lord Acton's famous "All power corrupts, absolute power corrupts absolutely" can be extended into the technological realm to recognize that technological power, when not constrained by institutional guardrails, inevitably corrupts the hands that hold it. By corruption, we understand both the subversion of larger objectives, whether pursuing international peace or the enlightenment of a community, to short-term, lesser objectives, whether individual enrichment, social control, or territorial conquest. Numerous examples abound. Nuclear weapons, as an example of absolute technological power, in 1945 ended the Second World War, yet for 18 years after that resulted in a nuclear stalemate first between the United States and the Soviet Union, later joined by France and Israel. Even today rogue nations such as Iran and North Korea threaten to destabilize the world by building a nuclear arsenal. The internet, created in the 1990s as a tool for bringing the world closer together, was soon discovered by "attention merchants" as a useful tool for harvesting the attention of billions of users for sale to advertisers, recruiting for dubious causes, and trade in child pornography. Even the automobile, perhaps an iconic technology of the twentieth century, led to the division of urban settlements as the wealthier were able to flee the cities for the suburbs.

As we have noted, institutions are the durable peace treaties that enable diverse communities and occupations to get along. Institutions are invested with a sense of identity—"who we are"—and with a cosmological horizon—where we fit into the universe. Human communities scan their horizon through institutions. Further, they learn to live together, although not always in harmony, through their institutional ecosystem.

In a rapidly changing world, the institutional ecosystem scrambles to keep up with the rapidly changing technology. Supply chains extending around the world challenge both the businesses that depend on them to satisfy their customers and the regulators that seek to assure product safety; social media, spanning continents is changing the meaning of "friendship." As rapidly changing technology introduces new threats and opportunities, social mores scramble to keep up. The changing of "friendship" from a personal attachment to a business opportunity illustrates the decadence of the technological society.

All of this is (or should be) obvious to any first-year sociology student, yet in today's technological society institutions are increasingly under assault. The technological society, in Ellul's account (Ellul, 1964), substitutes the logic or the machine for the logic of society, the logic of force, mass, and acceleration over the logic of compromise, nuance, and interpersonal adjustment, giving precedence to mechanical efficiency over social harmony. Compromise and interpersonal adjustment are more subtle than force, mass, and acceleration, yet privileging the technological values over the social values not only dumbs down society, but it also makes it more brittle and less capable of responding to unanticipated threats. Further, the decadence of the technological society, in which entertainment is substituted for engagement, and the promise of easy money undermines the work ethic that made America great.

Unanticipated threats are as old as empires, whether the barbarian hordes sacking Rome or the uprising of the Jacobins that toppled the Bourbon emperor Louis XVI. Democratic societies are more supple in responding to these threats, as many historical examples attest, yet in the technological society a new species of emperor, whether exemplified by Jack Welch or Elon Musk or Samuel Bankman-Fried, foreshadows collapse. Technological empires have a brittleness that earlier empires lacked, as we describe in the next chapter. The fall of the technological society probably will not be a gradual transition like the end of the Soviet empire but rather a dramatic collapse.

Each of these, whether cryptocurrency or space travel, has an institutional edifice, whether in the insurance companies that provide protection against risk, the laws that protect private and public property, the public agencies that promote and regulate trade and commerce, and, of course, the government. Money ($, €, ¥, £) is the medium of circulation within this orbit, commensurating the incommensurables of for example risk and government. Money *lubricates* the system; it does not *fuel* it. Money does not "make the world go 'round." Pumping more money into the system at best adds instability and more typically crashes the system entirely. This has been the story of many historical and contemporary bubbles we have mentioned.

In sum, in the technological society, new technologies add brittleness and instability as much as they add productivity, creating new but illusory robes of grandeur for a new species of emperor. The techno-titans, not unlike Nero in the final days of the Julio-Claudian dynasty in the first century BCE, are now presiding over a demonstrably collapsing civilization; the Dark Ages of the first millennium that follows may be the enjeu—what is at stake—of our technological civilization.

References

Bissell, T. (2019, January 29). An anti-Facebook manifesto, by an early Facebook investor. Review of Zucked: Waking up to the Facebook catastrophe. *New York Times.*

Crossman, R. H. (Ed.). (2001). *The god that failed.* Columbia University.

Ellul, J. (1964). *Technological society.* A. A. Knopf. (Original work published 1954).

Gelles, D. (2022, May 21). How Jack Welch's reign at G.E. gave us Elon Musk's Twitter feed. *New York Times.* (Adapted from Gelles, The Man Who Broke Capitalism).

Goffman, E. (1959). *The presentation of self in everyday Life.* Doubleday.

Immerwahr, D. (2019). *How to hide an empire: A history of the greater United States.* Farrar, Straus, and Giroux.

Koestler, A. (1968). The Act of Creation. In D. Lindsley & A. Lumsdaine (Ed.), *Brain function and learning, brain function volume IV* (pp. 327–346). University of California Press. https://doi.org/10.1525/9780520340176-014

McNamee, R. (2019). *Zucked: Waking up to the facebook.* Penguin.

Pfaffenberger, B. (1992). Technological dramas. *Science, Technology Human Values, 17*(3), 282–312. https://doi.org/10.1177/016224399201700302

Simmel, G. (1900/2011). *The philosophy of money.* Routledge.

Warren, Senator Elizabeth. (April 8, 2024). *Letter to the house financial services committee from Sen. Elizabeth Warren.*

Chapter 12
A More Brittle World

Abstract It is the thesis of this chapter, building on what we have observed regarding the character of technology, that this increasing vulnerability is not an unfortunate by-product, but rather inherent in the character of a technological society in which attachment and devotion to the machine has replaced attachment and devotion to living, breathing fellow citizens and life in general. The brittleness of the technological society, which many have noticed but not yet factored into its costs and benefits—its *externalities*—is an *institutional* choice, obscured by the short-term thinking of contemporary society, specifically in the West. We suggest the common technological thread of all of this is a tight coupling, situations where components in a system are highly dependent on each other. Facebook's algorithms do not give the user "here's another viewpoint to consider." Rather, the algorithms are tuned to elicit arousal, the strongest, usually negative reaction. We *know* how to create systems that are more reliable and less brittle yet have chosen as a matter of policy and commercial expedience to live in a more brittle vulnerable world.

We live in an increasingly brittle world. As many have documented over the past 40 years, our large-scale technological systems have made us increasingly vulnerable to both man-made and natural disasters. It is the thesis of this chapter, building on what we have observed regarding the character of technology, that this increasing vulnerability is not an unfortunate by-product, but rather inherent in the character of a technological society in which attachment and devotion to the machine has replaced attachment and devotion to living, breathing fellow citizens and life in general. The brittleness of the technological society, which many have noticed but not yet factored into its costs and benefits—its *externalities*—is an *institutional* choice, obscured by the short-term thinking of contemporary society, specifically in the West.

A. Batteau and C. Z. Miller, *Tools, Totems, and Totalities*,
https://doi.org/10.1007/978-981-97-8708-1_12

Normal Accidents in a Technological Society

The concept of normal accidents has been with us for nearly 40 years, yet few have grasped how ironic (and antithetical) it truly is. If something is "normal," can it be considered an "accident?" If I drive consistently on the wrong side of the road, is my inevitable head-on collision truly an "accident" or just inevitable, stupid, and expectable? In fact, "driving on the wrong side of the road" might be an apt metaphor for the increasing brittleness of a society that is dominated by hegemonic[1] technology. Building systems that are increasingly complex and tightly coupled is less a technological inevitability than it is a deliberate tempting of fate. On a country road with almost no traffic there is almost no hazard, but on a city street a collision is inevitable.

Charles Perrow, in *Normal Accidents* (Perrow, 1999), observed that when systems are tightly coupled and complex, accidents are expectable and require special measures to avoid. Perrow's analysis began with the meltdown of the Three Mile Island reactor in 1979, in which multiple systems, including confusing alarms, went off, leading to the meltdown of the reactor and a cascade of disasters. Perrow's concept has been applied to multiple technological systems, including air transport, financial institutions, and offshore oil drilling. In 1998, Long-Term Capital Management collapsed due to highly leveraged and opaque trades, creating a complex and tightly coupled financial structure that nearly wrecked the American economy. "Fintech," a concept that came into play only in the last twenty years, connects financial institutions, retail vendors, users' smart phones, and the Internet supposedly to link investments and payment processing more conveniently and seamlessly, but in fact add complexity and tight coupling[2] to a financial system that was already overburdened with payment processing. Fintech—financial technology—is the use of complex connections and algorithms to speed the flow of "money" including cryptocurrency (see Chap. 11) through the producing and consuming institutions of the society, in theory making it more productive but in fact making it more opaque[3] (Tett, 2010). Tight coupling often results in a system that is difficult to modify or scale because adjustments in one part necessitate changes in multiple other parts.

One example of an expectable (normal) accident was the Northeast Power Blackout of 1965. On the night of November 9, a relay tripped at the Robert Moses dam in upstate New York, causing a cascading set of failures that spread over eight

[1] Hegemonic systems here can be remarkably resilient. However, as we discuss in this chapter, they are not immune to brittleness and disruption. Internal contradictions, economic disparities, cultural shifts, globalization, technological advancements, and resistance movements can all contribute to the instability and eventual disruption of hegemonic power.

[2] Tight coupling, a term that comes from systems theory, refers to systems in which components are highly interdependent and closely linked. Characteristics of tightly coupled systems include high interdependence, fast response times, and minimal buffering and redundancy (Perrow, 1999).

[3] Gillian Tett, an award-winning journalist at the *Financial Times*, drew the ire of Wall Street leaders by warning of an unfolding financial crisis more than a year ahead of its arrival. Tett, who trained as an anthropologist, tells the unvarnished story of the people and events that were at the heart of the 2008 meltdown.

states plus the province of Ontario, leaving millions of citizens in the dark for more than 12 hours. There have always been local power failures, but the blackout of 1965 was notable for its geographic scope, a consequence of the scope of the power grid. In the years, since the publication *Normal Accidents* there have been numerous examples of large-scale disasters due to complexity and tight coupling. A 1982 book by Amory Lovins and Hunter Lovins, *Brittle Power*, rereleased after September 11, 2001 attacks, documents how America's energy system, both in terms of electrical grids, power generation, and even gas supply, has become more vulnerable to disruption, a vulnerability enabled in large part by technological developments. Large-scale power grids, like other optimized systems, are inherently vulnerable to natural or man-made disruption.

There are numerous examples of technological systems that are increasingly brittle. Perhaps at the top of the list would be supply chains, which in the last twenty years have proliferated around the world. Supply chains are, of course, a consequence of the increasingly global character of production, yet "just in time" supply, while minimizing costs of accumulated inventories, in fact increases brittleness. "Just in time" (JIT) contrasts to "just in case," the accumulation of buffers in warehouses to cope with unexpected demands or shortages. While "just in time" has demonstrably saved consumers, in the last two or three years in the aftermath of the COVID pandemic JIT has led to multiple shortages, whether of raw materials or semiconductors, or even household necessities and medical supplies. The brittleness of supply chains is perhaps an expectable consequence of engineering and optimizing an entire production organization, from raw materials (coal, iron ore) through intermediate goods to final assemblies. All this works within a factory, where product and process supervision and management can occur under one roof. However, imagining an entire continent or globe as a factory facing multiple cultural, linguistic, and geographic fractures demonstrably has not worked.

We noted the recent cause of global supply chains collapse which led to shortages in repair parts, semiconductors, and even toilet paper, was onset of the COVID pandemic in 2020. Pandemics are a consequence of a tightly coupled society, where people living in close quarters (i.e., cities) can quickly spread the contagion. Indeed, pandemics are as old as civilization, although have been on the upswing in the twentieth century. Notable pandemics include the Spanish Flu of 1918–1920, which killed more than a million people worldwide, the multiple SARS epidemics of 2003, 2009, and 2013, or the MERS epidemic of 2012. As global society has become more tightly coupled, pandemics have increased. Pandemics could be considered a "normal accident" in an increasingly globalized world.

Climate change is another example of a normal accident. Human societies have at various times and places exhausted their natural resources, whether the Norse settlements in Greenland between 985 and 1450 CE, on Easter Island (Diamond, 2005). An obsession with monumental statuary on Easter Island ultimately led to ecological collapse in the thirteenth century. During the Middle Ages, forests were sometimes over-cleared, and fields were exhausted by villagers that had not yet discovered the principles of crop rotation, but the possibility of exhausting global resources such as the atmosphere would have to wait until the Industrial Revolution.

Indeed, one could argue that the warming of the globe is a by-product not only of the industrial scaling up of production, but also of the fracturing of society. Governments, from the United States to Britain to China, know how to stem global warming by burning less fossil fuel, investing more in renewables, conserving energy, but such measures, notably in the United States, have become enmeshed in the politics of the "culture wars," a fracturing of society that we argue is a consequence of society's investments in technology.

A counterpoint to normal accidents theory was High Reliability Theory, developed by Karl Weick and Katheen Sutcliffe (2007), and applied to numerous high-risk systems, including aircraft carrier operations, air traffic control, nuclear power plant operations, hospitals, and other organizations. High Reliability Organizations (HROs) have multiple common features, including redundant resources, devolution of authority, pre-occupation with failure, reluctance to simplify, deference to expertise, and commitment to resilience. Redundant resources mean backup systems, so that if a primary system fails, the backup can take over. Devolution of authority places the ability of those on the front line to halt operations early, rather than watching the failure(s) cascade. On the flight deck of an aircraft carrier, even the lowest seaman can stop operations; he may have to explain why, but being on the front line he has greater visibility on impending failure. It is notable that all these examples of high reliability occur in military, government, or highly regulated environments. Left to itself, the technological society is inevitably brittle. It is also notable that in a fractured society, such quotidian matters as deference to expertise and commitment to resilience seem to be challenged (Weick & Sutcliffe, 2001).

A good illustration of the brittleness of a tightly coupled system is the 17,000 flights of Southwest Airlines that were canceled over the holiday season in 2022. A combination of tight scheduling and an unexpected winter storm forced flight cancelations all over the country stranding millions of travelers. Southwest, which had prided itself on smooth operations, suddenly found its operations disrupted by an unexpected event and had more canceled flights than other airlines affected by the same weather. For Southwest, the operational efficiency optimized by favorable weather turned into a major problem when confronted by unfavorable weather with long-term consequences for its reputation and brand.

Southwest's experience was foreshadowed by a development in freight railroads, Precision Scheduling Railroading (PSR), in which freight trains with hundreds of cars stretched for miles. Precision scheduling was implicated in a derailment in February, 2023, of a freight train near East Palestine, Ohio. Without getting into the details of freight operations, we can simply note that by optimizing the use of trains and train tracks, Precision Scheduling demonstrably subtracts from the overall resilience of freight operations.

Indeed, designing a society for high reliability, while not inconceivable, will remain mostly an academic exercise as long as members of the society feel "we have nothing in common." Healing the political (partisan, ideological, interest-group) and social (racial, class, regional) fractures in the society is the first step in designing a more resilient and robust society. As one anonymous lyricist stated, before we learn to weave, we must learn to mend.

Machinery of Deception

Perhaps the most extraordinary consequence of the technological society is the increasing tribalism of its politics, both in the United States and Europe. Much can be attributed to what Stuart Stevens characterizes as the "machinery of deception," the apparatus for sowing discord in contemporary discourse. Stuart Stevens, in *It Was All a Lie*, (2020) describes how the rise of right-wing media in the past forty years has created alternative realities that now fragment the body politic. The "Machinery of Deception" summarizes the rise of right-wing media, beginning with the periodical *Human Events* and William Buckley's *God and Man at Yale* (1951/2021), both published by Regnery publishing. This machinery is familiar, whether in the algorithms that Google and Facebook use for newsfeeds, or the lack of richness in contemporary online communication.

Perhaps the leading figure for industrial-strength deception was Donald Trump, who over his presidency made more than thirty thousand false or misleading claims. Perversely, Trump's supporters knowingly dismiss his falsehoods, because "he is doing it for us," thus feeding into and reinforcing the tribalism of contemporary society. Far less a "captain of industry" or a national leader than an entertainer, Trump is a master of deception, feeding the tribalism of contemporary society. He understood that in media there is an inevitable tradeoff between reach and richness: media with local reach, such as face-to-face conversation, have great potential for richness and depth, while those that reach around the world, such as text messages on the Internet, are impoverished, conveying less subtlety than a face-to-face conversation.

The common technological thread of all of this is a tight coupling. Facebook's algorithms do not give the user "here's another viewpoint to consider." Rather, the algorithms are tuned to elicit arousal, the strongest, usually negative reaction. We *know* how to create systems that are more reliable and less brittle yet have chosen as a matter of policy and commercial expedience to live in a more brittle vulnerable world.

These dangers are not simply confined to industrial systems but can (and do) embrace information systems as well. Another example of a brittle system is "reality TV." We place "reality" in quotes because in fact there is nothing real about it; a better term might be "unscripted" or "semi-scripted." "Reality" TV is very much a twenty-first century phenomenon, coincident with other matters such as the rise of cable TV and the rise of the Internet. Notable examples of reality TV include "Jersey Shore" (Salsano et al., 2009–2012) and "Keeping up with the Kardashians" (Seacrest et al., 2007–2021). When a reality TV entertainer with no notable political or business accomplishments became President of the United States, it represented the climax of a sorry chapter. There is a modest controversy over whether Donald Trump is truly a "successful businessman" or merely plays one on TV. Six bankruptcies, and financial losses in the billions, suggest that his "business success" is mostly performative.

Pandemics almost by definition tightly couple communities, and the only way to slow them down is through isolation. Yet when a pandemic strikes a community that already has many fissures (political, ideological, and cosmological), developing a collective response, which is essential, is harder than ever. These fissures are not simply fissures of differing political interests, but *cosmological* fissures, differences of "who are we," "how we live," and "what is our place in the universe." America leads the world in COVID deaths per million, not because Americans lack food and nutrition (although some do), but because America is becoming increasingly tribal, and mundane issues such as vaccination, mask-wearing, and social distancing have become signifiers of tribal identities.

There are many causes for this, not the least of which is the decline and corruption of the American empire. Just as Gibbon described in *The Decline and Fall of the Roman Empire* (1788/2003), increasing inequality and corruption led to collapse, so too in contemporary America increasing inequality and corruption are presaging a collapse of democracy. Foremost among these trends, as noted above, is the rise in the "machinery of deception." Stuart Stevens (2020), a long-time Republican operative until he became disgusted with how the Republican Party became Donald Trump, chronicled the rise of right-wing media at the end of the twentieth century and into the twenty-first. There is a self-reinforcing cycle here: The rise and proliferation of machinery of deception (most notably web pages) contributed to the fracturing of American society, not only in terms of partisan affiliations but also in basic sense of identity. Identity—the sense of "who we are" and "where we fit into the universe"—is fundamental to how we understand and make sense of the world around us. As we discussed in Chap. 8, "Who Are We?" one consequence of the technological society has been a splintering of identity, and hence less agreement on what is true and what we should do about it.

A technological and cybernetic perspective on the information ecosphere should make clear how technology promotes tribalism and social fracturing. Technology, almost by definition, lowers the cost of production, while it often lessens the quality of the product. In some cases, where there are established standards (such as those of the FDA for food safety) this is mostly harmless, although growing concern with "junk food" (rich in calories and saturated fats that related to negative health issue, but not toxic) has been a consequence of the industrialization of food production. However, where there is a lack of standards (other than the modest "content moderation" policies of social media) there is a rise in junk media, media what while not libelous (just as junk food is not poisonous) is certainly harmful to the health of public discourse, and in fact promotes social fracturing.

Junk media is central to the business model of TikTok, Facebook, Twitter, and other social media and cable networks. Social media posts are distributed based on sophisticated algorithms that select them in terms of their likelihood of evincing a (mostly negative) reaction or a pleasurable response that becomes addictive in nature. Junk media now predominates in the information ecosphere, feeding into a mindless numbing out and tribalism that, while always in the background of American politics, has now stepped into front and center stage. The fact that these platforms have (perhaps unwittingly) adopted junk media as central to their business models is indicative of a withdrawal from the public sphere into their own private glen.

Junk Media and Imagined Tribes

Benedict Anderson, in his book *Imagined Communities* (1983), describes how the invention of the Gutenberg Press in the 1450 s created the widespread dissemination of printed material and enabled hundreds of thousands of consociates, spread over entire continents, to *imagine* that they were sharing a conversation despite a lack of physical proximity. For the next four hundred or so years media advanced slowly, and the role of print media in contributing to multiple national revolutions, whether the American Revolution or the French Revolution, is well documented. In the twentieth century, multiple forms of new media, whether radio or television and now the Internet, have provided a foundation for new imagined communities, most notably the tribalism that proliferates on the Internet.

Tribes as a form of social organization stand in contrast to *nations*, which are fostered by shared narratives. As we discussed in previous chapters, in America, the narrative of conquering the wilderness and breaking away from the British Empire is a narrative that unites all Americans (or at least those of European ancestry), just as in France the narrative of Marianne at the barricades in 1789 unites the French nation. In China, the narrative of the Middle Kingdom, the source of harmony on earth, unites a nation extending to the far corners of Asia. By contrast, a tribe is preoccupied with its exclusive identity, typically revolving around a totem, a spirit being, sacred object, or symbol that serves to represent a group of people, a family, clan, or lineage. In preliterate tribal societies, totems were drawn from the natural world, and members of the eagle clan identified with the eagle, not so much in terms of biological descent as in terms of analogical resemblance, with the Eagle Clan being to eagles as the Wolf Clan was to wolves.

In a technological society, totems are primarily technological objects—guns, most notably, but also transportation devices and hand-held communications devices. The Harley-Davidson motorcycle pictured in Chap. 8, for example, nicknamed "The Hog," is a totem of masculinity in American culture, with an annual rally in Sturgis, South Dakota, and its own corporate-sponsored tribal assembly, the Harley Owners Group (HOG). Automobiles are signifiers of identity, including both gender, class, and occupation.

Similarly, in America personal firearms have *become* emblematic of tribal identities, not because the society has become physically more dangerous, but rather because many Americans have felt in recent years that their *identity* as an American was threatened. As noted in Chap. 3, gun sales boomed after Barack Obama was elected President in 2008, not because of any increase in threats to public safety, but simply because literally millions of Americans could not imagine that someone with a black skin could become President of *their* country. Guns are more accurately understood as *fetishes,* an object believed to have supernatural powers, rather than *tools* for self-defense.

With the rise of the Internet and junk media, the machinery of deception has switched into high gear. Ever since the Gutenberg press, mass media have had the task of sorting out fact from fiction, truth from falsehood. Newspapers in the colonial era were nakedly partisan, although the expense of operating a printing press and assembling an editorial team limited the number of news outlets. With broadcast media, first radio and then TV, the Federal Communications Commission established the Fairness Doctrine,[4] mandating that broadcast should provide an opportunity for rebuttal; Stations broadcasting controversial viewpoints had to give equal time to differing viewpoints, which often made them cautious of controversy. This was based in part on the fact that the airwaves are a common pool resource, shared by all. In the 1980s, Ronald Reagan abolished the Fairness Doctrine, which a Democratic-controlled Congress attempted to reinstate in 2011. With the rise of cable, the Internet, and Internet media *not* controlled by the Fairness Doctrine and in fact in some key quarters controlled by an alien (Rupert Murdoch, who changed his citizenship so that he could own TV stations in America) the media regressed to the earlier days of the penny-press, dominated by screaming headlines. The common technological thread of all this was a cheapening of discourse leading to a general fracturing of society.

Technological improvements arguably improve productivity, which is generally considered as a Good Thing. However, when the output is socially detrimental, whether in terms of junk media or pollution or carcinogens or other social hazards, then improvements in "productivity" are socially damaging. The automatic weaponization of technology in an increasingly hazardous world needs to be reconsidered.

Junk media is highly analogous to junk food, both in terms of its regulation (or lack thereof) and its harmful effects on the body politic. Just as junk food is not outlawed (although it is expected to have nutritional information as a warning), junk media, if it is not libelous, is legal, even if it is corrosive. Junk media creates and reinforces junk narratives, shared myths that increase tribalism in the society. This reinforces the junk politics that beset American society, the pursuit of power less in terms of any social or ideological agenda and more simply for the sake of having power. The failure of the Republican Party in 2020 to have a Party platform to stand on in their "shared" pursuit of national offices is perhaps the low point in American politics, yet for the many the whole point of winning an election was not to pursue any shared goals, but simply to exercise control and power to drive a specific agenda.

The corrosive effects of power for its own sake are well known. The next-to-final chapter in the decline of many empires, including the American Empire, is the accumulation of power and wealth not for any ideological or cultural agenda such as the advancement of Christianity, Islam, capitalism, authoritarianism or any other shared cause, but simply for imperial domination, for the grandeur of the emperors.

Junk relationships are analogous to junk money, money that is not backed up by institutional safeguards whether banks or the Federal Treasury. As described in

[4] The fairness doctrine of the United States Federal Communications Commission (FCC), introduced in 1949, was a policy that required the holders of broadcast licenses both to present controversial issues of public importance and to do so in a manner that fairly reflected differing viewpoints. In 1987, the FCC abolished the fairness doctrine, prompting some to urge its reintroduction through either Commission policy or congressional legislation. The FCC removed the rule that implemented the policy from the *Federal Register* in August 2011. https://en.wikipedia.org/wiki/Fairness_doctrine#cite_note-2 (accessed July 12, 2024).

the previous chapter, "cryptocurrency" is simply the latest in a long list of Ponzi schemes which, technologically enabled, provide a new avenue for the powerful and well informed to prey on the ill-informed. There is a long history in America of junk money, banknotes with little backing, particularly on the frontier, swindling the farmers. Much of our contemporary politics might be considered as a pyramid scheme in which those at the top, by pretending to be champions of the people are in fact exploiting their base. Similarly to the analysis we presented in the previous chapter of how the *value* of money has an institutional edifice in banks, the state, trade and commerce, tangible goods and services, as well as risk-mitigation institutions and strategies. So, too, media require an institutional edifice in the professionalism, training, ethics, and standards of reporters and editors. This institutional edifice includes both corporate and family ownership of media properties such as the Washington Post and the New York Times, and the academic programs that students learn the craft of journalism.

"Institutions" may sound dry and sociological, but in fact are the heart and soul of a society. Institutions embody several key aspects of a society, including identity ("who are we?", see Chap. 8), and cultural resources of kinship and sacrality. When a major media property, Fox News, owned by a foreign national, decides to debase journalistic standards in order to reap the profits and maintain the loyalty of its audience, it was a low point not only in journalism but also in American history. A leading organ of the "Fourth Estate" had surrendered, not so much to a foreign power as to foreign influence.

Finally, institutions have a poetry that inspires, whether within their halls or within their imagination. The poet-librarian, Philip Larkin's "Library Ode," celebrates the library as a place where the old and new coexist, with new eyes finding old books and old eyes renewing with new books, symbolizing the connection between youth and age.

New eyes each year
Find old books here,
And new books, too,
Old eyes renew;
So youth and age
Like ink and page
In this house join,
Minting new coin.

Similarly, museums are not simply warehouses for historic relics, but temples to a shared heritage, as exemplified by W. H. Auden's 1939 poem "Musee des Beaux Arts" (1940).

About suffering they were never wrong,
The Old Masters: how well they understood
Its human position; how it takes place
While someone else is eating or opening a window or just walking dully along.

How, when the aged are reverently, passionately waiting
For the miraculous birth, there always must be
Children who did not specially want it to happen, skating
On a pond at the edge of the wood:
They never forgot
That even the dreadful martyrdom must run its course
Anyhow in a corner, some untidy spot
Where the dogs go on with their doggy life and the torturer's horse
Scratches its innocent behind on a tree.

Junk relationships are not so much openly predatory in the manner of tenant farmers or under-paid industrial workers, as they are simply transactional, motivated by a "what's in it for me?" sentiment, rather than "how can we prosper together?" George Foster's article, "Peasant Society and the Image of Limited Good" (1965) described a pre-industrial counterpoint to the "culture of poverty" in which people take a basically selfish approach, limiting their ability to build more cooperative generalized reciprocity communities. Although this article, like the idea of the "culture of poverty," created considerable controversy, with many questioning their relationship to colonial and class domination, it is unquestionable but that the technological society has created a new form of peasant society, only now not tied to the land and agriculture, and more tied to the machine and requiring increasing amounts of (human and industrial) capital to master.

The summarize; by placing social relationships at the mercy of the machine, the increasing material affluence is *not* offset by an increase in social harmony or shared objectives, and in the long run is not sustainable. The fracturing of not only American society, but of other nations where the Internet is minimally regulated, is testimony to this. The information ecosphere, when not surrounded by institutions that have a regulatory mandate, as described in the previous chapter, increases and tribalism and the polarization of the society.

Even though the point may seem obvious, it is worth reflecting on the distinction between transactional and other types of relationships. Kinship, a statement we *share* something, whether ancestry or other element of identity, is *not* transactional. Kinship is not embraced in "the art of the deal." By contrast, a transactional relationship, a "what's in it for me?" relationship, is a denial that we have anything in common other than an exchange. The art of kinship is the art of sharing, which often requires substantial nuance. By contrast, the art of the transaction is the *calculation* of how much I need to give up in order to get something that I want, and is it worth it to me. Transactional relationships are the foundation of business, but they are *not* a foundation of society.

A social fabric is made up of multiple types of relationships, some voluntary, some coercive, some altruistic, and some selfish. The *art* of building a resilient social fabric comes from balancing multiple relationships and advantages for the greater good. America is fortunate in that it is not dominated by predatory relationships, although in some quarters, most notably low-income neighborhoods and the pinnacles of finance, predation IS the rule, not the exception. Transactional relationships—the art of the deal—make for a more prosperous society, even if that prosperity is brittle and unevenly distributed.

Sharing—the art of the gift—creates a fellow-feeling, a resilience that can outlast the ups and downs of business cycles and pandemics, and even political collapse. Exchanges that on the surface are voluntary "gifts" can be transactional, predatory, or kinship-creating. Perhaps the least obvious is predatory-gift-giving, the overly generous gift to induce a sense of guilt on the part of the recipient, or public acclamation for the giver.

Marcel Mauss, in *Essai sur le Don* (translated as "The Gift: Forms and Functions of Exchange in Archaic Societies"), argued that gift-giving entailed three obligations: the obligation to give, the obligation to receive, and the obligation to reciprocate. Gift-giving, for Mauss, was *not* transactional, but rather a matter of open-ended generosity, and a foundation for society. To sum up this section, a resilient, non-brittle society must *cultivate* the art of generosity, not only among family members but within communities and in fact the entire polity. Generosity, as Jesus observed, is truly anonymous, hidden, not expecting reciprocal payment or publicity.

> Therefore, when thou doest thine alms, do not sound a
> trumpet before thee, as the hypocrites do in the
> synagogues and in the streets, that they may have glory
> of men. Verily I say unto you, They have their reward.
> (The Holy Bible, King James Version, Matthew 6:2)

Modern technology has supplied trumpets for publicity, but not for generosity.

Conclusion

Although we cannot posit a technological cause for all these faults and fractures, the hegemonic nature of modern technology demonstrably gives them force and vigor that would otherwise be lacking. Attitudes become hardened, and when reflected in governing decisions, whether the appointment of Supreme Court justices or the passage of laws restricting voting, the net result is a more brittle, less resilient society, less capable of adaptation, and coping with future threats whether natural, industrial, or global. Perhaps the biggest threats are *predatory*, the preying of the strong, the well-informed and well-resourced, the well-connected on the weak, leading to an increasingly fractured society.

In the mid-twentieth century, America distinguished itself by leading the world (not perfectly), whether through the foundation of the United Nations, through the Marshall plan for rebuilding Europe announced in 1944, through its role in establishing NATO, or through the Organization of American States founded in 1948 for building up the Americas. The Civil Rights Movement was also part and parcel of this leadership. The prosperity of the United States in the mid-century was an expectable outcome of this leadership.

The "modest interventions" that anthropologist and designer Lucy Suchman calls for (discussed in Chap. 5) would provide a welcome alternative to the massive change that technological innovations often provide, creating more vulnerability in systems and a brittle world. As we explore in the following chapters, designers may be able to teach technologists to think more modestly, less massively, in developing systems and institutions for the world of the twenty-first century.

The brittleness of contemporary society should alarm anyone concerned with the world beyond the here and now. "How will you and your community be remembered in years to come?" is a question that should occupy all but the most self-absorbed, narcissistic personalities. The fact that American society is becoming more self-absorbed and narcissistic, at least in its highest quarters, is the tragedy of the hegemonic technological society.

References

Anderson, B. (1983). *Imagined communities: Reflections on the origin and spread of nationalism.* Verso.

Auden, W. H. (1940). *Another time.* Random House.

Buckley, W. (2021). *God and man at Yale: The superstitions of 'Academic Freedom'.* Regnery Gateway.

Diamond, J. (2005). *Guns, germs, and steel: The fate of human societies.* Norton.

Foster, G. M. (1965). Peasant society and the image of limited good. *American Anthropologist, 67*(2), 293–315.

Gibbon, E. (1788/2023). *The decline and fall of the Roman empire.* Legare Street Press.

Larkin, P. *My poetic side: Library ode.* https://mypoeticside.com/show-classic-poem-1608

Lovins, A. B., & Lovins, L. H. (1982/2001). *Brittle power: Energy strategy for national security.* Brick House Publishing Co.

Mauss, M. (1954). The gift: Forms and functions of exchange in archaic societies (I. Cunnison, Trans.). Cohen & West.

Perrow, C. (1999b). *Normal accidents: Living with high-risk technologies.* Princeton University Press.

Salsano, S., Jeffress, S., & French, J. (2009–2012). *Jersey Shore* (MTV).

Seacrest, R., et al. (2007–2021). *Keeping up with the Kardashians.* In. Los Angeles, CA: E!.

Stevens, S. (2020). *It Was All a Lie: How the Republican party became Donald Trump.* Vintage.

Tett, G. (2010). *Fool's Gold: How the bold dream of a small tribe at J.P. Morgan was corrupted by Wall Street and unleashed a catastrophe.* Simon & Shuster.

The Holy Bible, King James Version, Matthew 6:2

Weick, K. E. & Sutcliffe, K. M. (2001). Managing the Unexpected—Assuring High Performance in an Age of Complexity. San Francisco, CA, USA: Jossey-Bass, pp. 10–17. ISBN: 978-0-7879-5627-1

Weick, K. E., & Sutcliffe, K. M. (2007). *Managing the unexpected: Resilient performance in and age of uncertainty* (2nd ed.). Jossey-Bass.

Fairness Doctrine (Federal Trade Commission). (1949–1987). Wikipedia.

Chapter 13
The Enchantment of Technology

Abstract In this chapter, we examine the cultural roots of our love affair with technology. Drawing on Marshall Berman's insights in *All that is Solid Melts into Air* (1988), we examine the close relationships among technology, modernity, and modernism. Technology, we suggest, is as much about *images* as it is about functionality, and the chief image or style associated with technology is modernism, an artistic movement that came into being only in the early twentieth century. The "Metaverse"—Facebook's construction of its new virtual reality—is the perfect representation of this love affair with technology.

The *magic* of technology, its capability to do miraculous things and enchant our senses is part of the allure of the technological society. In this chapter, we explore the enchantment of technology.[1] Much like the sacred aura of priests or the grandeur of emperors, the magic of technology is central to the imperative order of a society ruled by a hegemonic technology, just like the miter and scepter were central to the order of earlier societies. As we uncover this *hidden* enchantment, we will uncover the boundaries of a technological society.

A Generalized Theory of Magic

Almost by definition, magic is a residual category, embracing those phenomena that we can superficially observe but not explain. The imbalance between observation and explanation is central to the concept of magic. Entertainers (magicians) use sleight-of-hand to perform magical tricks, although the audience understands that this is

[1] In what now seems like a technological Age of Innocence, MIT Media Lab scientist David Rose presented a "vision of the future, [in which] technology atomizes, combining itself with the objects that make up the very fabric of daily living. Such innovations will be woven into the background of our environment, enhancing human relationships, channeling desires for omniscience, long life, and creative expression. The enchanted objects of fairy tales and science fiction will enter real life." (2015).

A. Batteau and C. Z. Miller, *Tools, Totems, and Totalities*,
https://doi.org/10.1007/978-981-97-8708-1_13

an illusion. Contemporary forms of magic, on the other hand, such as advertising, rely on the observers being taken in by the illusion: advertising persuades millions that they can become more attractive, gain friends and influence, or improve their health simply by purchasing the right products. Although psychologists can explain how this hidden persuasion works, for most of us it is just magic. Contemporary magic, in contrast to the traditional stage performances of magicians, have become central to the institutions of the post-industrial society, a difference in practice from pre-industrial magic.

This disconnect between observation and (hidden) dynamics is at the heart of magic. Whether the dynamics are mechanical or informational or interpersonal, millions of *observers* cannot explain why the airplane flies or why the spreadsheet tabulates balances or why the drug elevates their moods or why an orator sways their allegiances—it is magic. Deep down, they know that some mechanism is at work, even if they do not understand the mechanism. Closely connected with this invisibility is the *enchantment* of magic, compelling attention even if one cannot explain why or what is going on. In fact, excessive attention to the mechanism of what is going on would detract from the magic.

Part of this enchantment is the mysterious power of technologies. When people understand power, whether the overt power of an army or a CEO, they can *rationally* adjust to it, whether through submission, negotiation, manipulation, resistance, or cooperation. When the power is opaque and mysterious, a different set of attitudes sets in. If, somehow, they are convinced that the power is rational and mysterious, another set of attitudes comes into play.

In 1957, at the height of post-war prosperity, Vance Packard published *The Hidden Persuaders,* a critique of the advertising industry. In *The Hidden Persuaders* (1959) Packard explored how advertisers, using sophisticated psychological techniques, manipulated consumers toward their products, or more generally we might add toward the idea that consumption in general is the key to personal fulfillment. Some of the psychological impulses that Packard identified included emotional security, sense of power, and immortality, all of which for the average consumer, are only magically achievable. The idea that these deep-rooted psychological needs can be attained through consumption (arguably but one corner of human existence, alongside friendship and family, religious worship, and political loyalty) is, well, simply magical. Consumption, although essential for biological existence, is but one small corner of *social* existence, alongside friendship and kinship.

Magic disconnects subjectivity and agency: *subjectivity,* central to all perception, and *agency,* one's capability to cause effects, have been central parts of human existence ever since *homo habilis* emerged in Africa. As we explored in Chapter 7, "The Narratives of the Machine", human culture and cultures spring from human capabilities including perception, cognition, and interpersonal connection, not simply the use of tools.

Another phenomenon that is perceived as magic is the wondrous achievements of technology. Again, engineers can explain the physical dynamics of flight or the processing of megabytes of data, but for the average user with his cell phone the fact that he can communicate around the world in real time with up to 1000 "friends" on any continent, while seated at the kitchen table, is just magic. Interest in magic shot up in the last 30 years as the Internet took off; although specialists can explain the workings of the Internet, for most users, it is simply magical.

Despite living in a rational age, most people have a hankering for magic. This is easily demonstrable in the appeal of advertising, an industry with nearly $140 billion in revenue in 2022 in the U.S. alone.[2] In comparison, the residential construction industry, which arguably adds significant elements to a fulfilling life, spent $404 billion.[3] Similarly, people look for techno-fixes for numerous everyday afflictions, whether social isolation, an escape provided through social media, ignorance (a problem solved with search engines that do not always curate truth from falsehood), or social status, a problem solved with what David Hess (1995) in *Science and Technology in a Multicultural World* has called techno-totems. Techno-totems are technological objects such as cars and smart phones that signal *social* identities. Not only is most of the population unable to explain how they work but many are also unaware of how they distort their desires. Like a magician pulling a rabbit out of a hat the magic works because of a willing and often eager suspension of disbelief, or perhaps an intense willingness to believe in technology as the proverbial silver bullet.

This willing suspension of belief stems from the fact that techno-fixes and technological explanations are *simpler* than the messiness of social interactions and human emotions. In any human community, there are a myriad of moving parts—individuals, groups, emotions, attentions—that are far more complicated than any cybernetic device. It is the limit (or laziness?) of the human mind rather than the inherent character of the device that attracts people to techno-fixes rather than what we might call empathy-fixes.

Part of the allure of magic today is a desire for enchantment, a taking leave of one's senses even for momentary rapture. In a rational society enchantment has little place, outside of romantic relationships and possibly political cults, but in numerous typically public situations, whether on stage, or the movies or in religious revivals, *enchantment* is supplied by magical means, whether that of an entertainer or communication with the God. Enchantment is a rapturous lost-in-the-moment situation, a freeing from worldly cares, even if only in a liminal space. Liminality, "betwixt and between," is a central part of the dynamism of any society. A society or a world without liminal spaces and episodes is a static, frozen in space and time.

Perhaps the ultimate in magic today is social media, through which one can maintain a network of up to 5000 "friends" (the arbitrary limit imposed by Facebook) without leaving one's home.[4]Facebook in fact collects terabytes of data on its users

---

[2] In 2022, the global advertising industry generated significant revenue, with the digital advertising sector alone reaching $139.8 billion in the United States. (Source: Internet Advertising Revenue Report, April 2023).

https://www.iab.com/wp-content/uploads/2023/04/IAB_PwC_Internet_Advertising_Revenue_Report_2022.pdf

[3] Source: Mordor Intelligence https://www.mordorintelligence.com/industry-reports/us-residential-construction-market

[4] Refer to the discussion in Chap. 11 (Fig. 11.2 Dunbar's number).

(now designated by Mark Zuckerberg as "people"), for purposes of selecting targets for advertising using algorithms that defy our comprehension. Most users ("people") do not understand what an algorithm is or how it works to manipulate their attention. Everyone who uses social media has had the experience of a pop-up ad closely related to their recent browsing, purchasing history, or even what they thought was a private conversation. A pop-up ad is no less magical than a rabbit pulled out of a hat. We might call these "magical friends" because one rarely has a face-to-face conversation with them and rarely knows their capabilities, or, in fact, if they are real people or bots. The *enchantment* of social media, the fact that it absorbs multiple human drives and desires with greater immediacy than any effort to make face-to-face connections with living, breathing, non-bot friends. These desires, which Rose, in *Enchanted Objects* (2015) are fundamental and universal, include omniscience, telepathy, safekeeping, immortality, teleportation, and expression. For example, a Facebook chat with a "friend" or a Zoom session with a colleague in Africa is teleportation, just magic.

This disconnect between visual observation and personal connection is evident many aspects of contemporary life. In small-scale village communities, where face-to-face relationships predominate, this is not an issue, but in contemporary society there are many examples of others impacting one's life out of visual range. This is so ordinary that we scarcely think about it, except when observation intrudes into personal lives, whether in the surveillance state, notice of a hacked account, or the algorithms tracking of one's browsing on social media.

The "surveillance state" is very much a twenty-first century phenomenon, enabled by the Internet and the fact that much of social life and commercial activity has migrated to the Internet. Whether the tracking is in the hands of state actors or commercial actors (increasingly an irrelevant distinction, since commercial actors and entertainers often have greater power than the state) in all cases *someone* is watching, someone out of visual range: a one-way capability.

Surveillance is nefarious black magic and can be fine-tuned to numerous personal characteristics, including race, gender, education, occupation, shopping history, and even residence. Previous examples of surveillance states included Nazi Germany and Soviet Russia, whose primitive tools were not sufficiently subtle to track citizens' whims. Today surveillance is contained in every web browser and Internet search, computing multiple personal characteristics from browsing history. The "New Jim Code," in Ruha Benjamin's words (2019), reinforces the racial divisions that have beset America at least since the seventeenth century, only now maintained with mouse-clicks and browsing history. In the nineteenth and twentieth centuries, Jim Crow laws denied housing and education to citizens based on the color of their skin. Such overt discrimination was outlawed by the court decisions and civil rights laws of the 1960s, only to be replaced by more subtle discrimination based on inferences of race through shopping history and employment history. The Internet has replaced Jim Crow *laws* with Jim Crow *algorithms,* effectively maintaining the second-class status of those who fall on the wrong side of the algorithm. Black magic is a dark form of nefarious enchantment, horrifying instead of enchanting.

Magic and Enchantment in Contemporary Society

We lump these phenomena with "magic" because in all cases there is a disconnect between visual observation and interpersonal action, sometimes benign, but with an all-powerful state such as China or Russia the effect is malign. This disconnection, or perhaps alienation, is an individual isolation that is endemic in contemporary society. *Social* ties, the substance of society, are pulverized into tweets and screenshots. However, states are perhaps not the dominant actors here: social media have all-seeing eyes that track mouse-clicks and, using sophisticated algorithms. As noted, many of us have had the experience of a pop-up ad closely matched to their recent browsing history. Is this a sophisticated manipulation of one's attention?

James Howard Kunstler, in *Too Much Magic* (2012), described how contemporary society, with looming exhaustion of petroleum reserves and the specter of climate change and over-population, has continued to maintain a trajectory of car-addiction ("Happy Motoring") and suburban sprawl as though energy reserves and landscapes and suburban tract housing are inexhaustible. (Armstrong et al, 2018; Chomsky, 2009; McKibben, 2008). Debates about climate change and exhaustible resources go back at least to the 1972 report for the Club of Rome, *The Limits of Growth* (Meadows, 1972), which suggested that the Earth's resources were not inexhaustible. The fact that these issues are still being debated, a half-century later, suggests that optimism over unlimited natural resources is endemic to a society in which technology is dominant.

In contrast to the down-to-earth seriousness of the seventeenth century Puritans or the nineteenth century Victorians, the ongoing addiction to a demonstrably unsustainable way of life packaged in big box department stores, shopping malls, and a seemingly endless number of online sites. Different ages have their own architectural achievement, whether the castles of the ancients or the cathedrals of Middle Ages. In the modern era, the notable architectural achievements begin with the factory and the office building and the train station, and in the twentieth century expanded to include the shopping mall. The enchantment of shopping in a mall or a "big box" is best understood as magical thinking. These twentieth century cathedrals have replaced worship with shopping, spiritual attainment with material consumption.

Similarly, Alfred Gell, in "Technology and Magic" (1988), asserts that "Technology is inadequately understood if it is simply identified with tool-using. Gell describes instead three domains of technology: technology of production (the customary understanding), technology of reproduction (such as kinship), and "technology of enchantment." This final domain includes music, rhetoric, dances, gifts through which members of a society *enchant*—psychologically manipulate—their consociates. The hidden connection between cause and effect, and the enchantment of hidden effects, is a pursuit of which is a uniquely human capability. In Gell's words,

> The technology of enchantment is the most sophisticated that we possess. Under this heading I place all those technical strategies, especially art, music, dances, rhetoric, gifts, etc., which human beings employ in order to secure the acquiescence of other people in their intentions

> or projects. These technical strategies – which are, of course, practised reciprocally – exploit innate or derived psychological biases so as to enchant the other person and cause him/her to perceive social reality in a way favourable to the social interests of the enchanter. (1988, p. 7)

Gell's optimism on reciprocity might be questioned when one of the actors is an all-powerful corporate entity or state manipulating the lives of its citizens. Everyone, other than the most sober Puritans, welcomes (at least in measured doses) enchantment, whether with a lover or with a deity. Enchantment is far more than physical or intellectual pleasure, having a depth and soulful resonance far beyond the here and now.

The theme of "too much magic" applies not simply to miraculous phenomena such as energy consumption and real estate, but also to quotidian events such as pop-up ads and facial recognition technologies. In all cases, there is a disconnect between the social impact of events and their visibility to ordinary (i.e., non-corporate and non-state) citizens. The net result is that the magic of technology concentrates more and more power in fewer hands, and the *invisibility* of these hands to the less powerful is part of their durability.

Out of this ash heap of bots and trolls, we may recover our lost sense of agency. *Agency* is not simply cause-and-effect; rather, in the vernacular of blues, gospel, and soul music, it is "call and response," a *social* and not simply physical activity. Wandering out of the thicket of social media we can rediscover the authentic value of personal connections, multi-channel relationships where voice, visibility, shared interests, and shared memories count for more than anger and outrage (Stuckey, 2013a, 2013b). Shared experiences, authentic connections, and an authentic sense of self in place of the illusions of power and grandiosity are more the *substance* of society than evanescent memes and screenshots. A sense of community, of shared interests beyond the immediate household and kindred.

The "eclipse of community," which many have remarked on (Stein, 2015), a concept that took off in the 1950s, was a development aided by technology (notably the Industrial Revolution). This division besetting not just American but many "advanced" societies, notably post-industrial or technological societies. In contrast to the "community studies" that marked American sociology for the first part of the twentieth century, whether Warner and Lunt's *Yankee City* (1963) series or Park and Burgess's *The City* (1925), we now have what might be called "alienation studies." Foremost among these was Reisman, Glazer, and Denney's *The Lonely Crowd* (1950), which examined modern society in terms of the trajectory from tradition-directed to inner-directed (the society of the "protestant ethic" Weber 1930) to other-directed, in which people yearn for connection.

Community studies were a dominant trend for most of twentieth-century sociology exemplified by Robert and Helen Merrell Lynd's *Middletown* (1929). Other works in this genre coming out of the Chicago School included and Davis and Gardner's *Deep South* (1941). These studies examined locally defined aggregations disconnected from larger national or international contexts, ironically in a century in which the global reach of multi-national corporations and technologies was on the rise. A counterpoint to community studies was what we might call "alienation studies," beginning with Erich Fromm's *Escape from Freedom*, (1941), continued with Riesman, et al. *The Lonely Crowd* (1950) summarized in Melvin Seeman's article in the 1975 *Annual Review of Sociology* (Seeman, 1975). This tension or interplay between community and alienation seems to be a dominant theme of industrial and post-industrial society.

As we have argued, "social" media we must make clear is *anti-social* media, grinding down a society into eyeballs and mouse-clicks, manipulating users ("people") into constituent elements of attention and transaction. For those who harvest billions of advertising revenue, this is a feature, not a bug. Focused only on private accumulation rather than any sense of a common good, this clearly is not a problem, but for the growing numbers who are becoming *disillusioned* with life on the screen, this may mark the opening of an opportunity to design, build, and create a society in which the hegemonic technology is not the rule.

The Magical Town Square

Perhaps the ultimate magic today is the new town square, Twitter (X), where literally billions of users every day gather to exchange ideas, provocations, experiences, and political preferences, using only text messages (no other sensory inputs) all while sitting at the office or remote working from home. Unlike the traditional town square or crowded auditorium of face-to-face interaction, on Twitter you can exchange messages with … bots? Trolls? Provocateurs halfway around the world? Long-lost acquaintances? Future lovers? Chat GPT? On Twitter, limited to 280 characters, communication is inevitably superficial, and in fact curated for its potential for emotional arousal. The *business* of Twitter is in fact harvesting *attention*, not participation, giving it revenues of more than $1.5 billion per year.[5] The *magic* of Twitter (X) is in its creation of an illusion of free marketplace of ideas. Further, one can argue the *business* of Twitter under its new owner is sowing chaos. Its growth over the past 20 years closely tracks the increasingly chaotic character of politics not only in America but in many countries around the world.

On the other hand, the *genius* of Elon Musk is that he understands the disruptive dynamics of this new medium far better than old-media pundits. Just as Adam Smith identified the wealth of nations stemming from free markets (rather than the plunder of colonies), so too the wealth of cyberspace stems from mouse-clicks and screen shots, and the *traffic* that these generate. We await a comprehensive accounting of the traffic of URLs. "Influencers" count for more than merchants and factory owners in the economy of the Internet, and the regulation of privacy in the face of intrusive search engines is another chapter waiting to be written. Just as the Food and Drug

[5] Twitter has experienced significant revenue changes since Musk took over the company in October 2022. In 2022, Twitter's annual revenue was $5 billion. After the acquisition, advertising revenue saw a dramatic decline, with reports indicating an 89% drop by March 2023 (https://www.thewrap.com/twitter-ad-revenue-plunges-elon-musk-bloomberg/ accessed July 13, 2024).

Administration intrudes on factory production of food, so too the yet-to-arrive Phone and Screen Administration will intrude on the economy of the Internet.

We do not intend to imply any causation here. Rather, Twitter's fragmentation and chaos are part of a much larger social breakdown of the technological society, in which respect for institutions is replaced by fascination with screen shots. In the technological society, as Ellul (1954/1964) and many since him have identified, humanity becomes more preoccupied with means rather than ultimate purposes.

Conceptually, the "town square" is a *public* space, but in fact this new magical town square is a private domain, now owned by the world's richest men. Elon Musk whose erratic behavior creates chaos on the platform makes the rules and moderates the content, causing advertisers (the principal source of revenue) to flee. Elon Musk exhibits all the erratic and chaotic behavior associated with historical emperors, whether Caligula, with his ambitious construction projects, or Henry VIII and his many wives, both proclaimed their grandeur, and like both Caligula and Henry, the chaos he creates is worn proudly.

In this new magical town square, one can learn new things both true and false, make new friends (including bots and trolls), air grievances and celebrations (real and imagined), explore distant lands around the globe or out in space from literally anywhere. The superficiality of this new form of social life, and the fact that it crowds out other forms of interaction impoverishes society.

Unlike the traditional town square, where neighbors and fellow citizens could meet face-to-face and exchange thoughts and opinions, adjusting their acceptance of thoughts in terms of their previous acquaintances, and signal their reactions with facial cues and body language, in this magical town square the channels of communication are limited and sometimes obscure. A key issue with this new town square is, "Who makes the rules? Who moderates the discussion, discarding provocative tweets and highlighting acceptable content?" The answer is the firm, owner with oversight (technically) provided by the Board of Directors. Unlike the "eclipse of community," noted above, corporate rather than communal, collective control is perhaps the ultimate achievement of the technological society.

To sum up this discussion, the *magic* of Twitter (X) is that it can sow chaos *invisibly,* and in fact its business model is based on this, making a business of chaos. The degeneration of society into dozens if not hundreds of brawling tweeters is good for Twitter's advertising revenues, at least it was until advertisers realized it had degenerated into an environment that was not productive and was instead potentially damaging their brands.

While designating Twitter as the new town square may *sound* democratic, in fact it is not just despotic but potentially totalitarian, because users can never really know who is moderating and curating their messages. It can be argued that a saving grace is the limited attention span of the Chief Twit and his chaotic management style, exemplified by his first day at (then still) Twitter headquarters in 2022, in which he arrived carrying a kitchen sink, (see above Chapter 11), but the trail that he is blazing for other social media (for example, "Truth Social") will beckon until societies begin to understand that junk media (see above, Chapter 12) is no less harmful than junk food.

Appeals to "free speech" are little more than a misdirection: "speech" is never "free," whether in a worship service or in a library or at the family dinner table. The recognition that there *should* be limits on online speech is fundamental to civility, a value that is increasingly lost in advertising-driven social media and cult-driven politics or celebrity-driven entertainment. Civility, the respectful interaction of citizens whether in the town square or at the family dinner table is increasingly a lost value in the technological society.

Civility is, in fact not a *public* good but a *club* good, *shared* by other members of the civilization. Proper manners and respectful behavior are tokens that one belongs, whereas *un-civil* behavior, even if not forbidden by statute, marks one as an outsider. Civility takes on a different aspect in Western and Eastern civilizations, whether in terms of gender roles or public appearances. Not unlike the Visigoths that sacked Rome in the fifth century CE, the twits of anti-social media are now sacking supposedly "civilized" societies. Unlike the Visigoths, whose weapons were only clubs and spears, today's twits are using the most up-to-date technologies to sack the town square.

The Re-emergence of Society

As the impoverished nature of social media becomes increasingly apparent, in a society in which technology is a convivial actor rather than a hegemonic tyrant we may rediscover the positive values of society and connection, as contrasted to individual values of consumption and gratification. As we have argued, *connection*, the essence of society, is far more than transactional: Connection is a good-in-itself. It is not reducible to gratification or accumulation or domination. All of these, of course, are present in society, but a society that was built solely on gratification or accumulation or domination would be fundamentally impoverished. By contrast, *sharing*—of goods, spaces, experiences, and identities—creates a more fulfilling social life.

Society is built on multiple connections, multiple sensory channels (visual, tactile, auditory, even olfactory), and multiple shared memories that make one's consociates *real*, that is more than simply words and images on a screen. It is based on *sharing* spaces (neighborhoods), family members (kinship groups), institutions (structures), resources (industries and enterprises), political authority (parties), and even visions (religious congregations). A society that has not learned how to share is no society. As obvious as this is (or should be), millions of users ("people") are on the precipice of plunging into Mark Zuckerberg's "metaverse," described in Chap. 10, an immersive "reality" that substitutes artificial-intelligence simulations for living, breathing people, places, and things. As obviously nonsensical as this is, technological titans are investing *billions* of dollars in creating this artificial "reality," which can only crash and burn, much like the toxic street fight where nothing is too outrageous or offensive that Musk's Twitter (X) is spiralling into.

An important part of connection for any healthy society is *trust*, the multi-valent, open-ended assumption that the other party is not going to harm me, or at least has my interests at heart. As reported in the 2023 "Edelman Trust Barometer".[6]

> Trust has declined around the world, for reasons far too complex to explore here. We can summarize to note that the media in general, and the internet in particular, far from "bringing the world closer together," in fact are major forces in creating tribal divisions both around the world and within industrialized countries.

Sharing is a basic human propensity, creating a sense of "who we are." At a village inn, or around the family dinner table, or in a bedroom, people use sharing to build connections and identities. When sharing happens online, often at the expense of face-to-face it signals a crumbling of sociality, the degree to which social groups form cooperative societies. Sociality includes the development of social norms, institutions, and cultures that facilitate living in communities and cooperating with one another.

Face-to-face experiences have a richness and complexity that screen-to-face experiences can never duplicate. "Life on the screen" (Turkle, 1995) becomes a new definition of poverty, not simply poverty of private goods but poverty of all the other goods (public goods, club goods, common pool resources) that make up a fulfilling society. In contrast to a balanced portfolio of goods and values, the technological society offers only an impoverished bushel of private experiences. While this may be acceptable to those whose business is harvesting attention, as Tim Wu describes in *The Attention Merchants* (Wu 2016) for the mass of citizens it is a new road to serfdom. The "attention merchants" of the technological society have perfected the business model of manipulating, aggregating, and packaging and selling the attention of billions of pairs of eyeballs around the world as the foundation of a consumer-driven economy.

Friedrich Hayek, in *The Road to Serfdom* (1944) argued that the central planning exemplified by Soviet Russia, created new forms of unfreedom. Fifty years later the Soviet Union came crashing down. Similarly, we can expect the empires of Facebook and X (formerly Twitter) to come crashing down, although what will fill the void remains to be seen.

Empires, not unlike Ponzi schemes (see Chap. 11 on cryptocurrency) amass power and wealth by convincing their investors of the inevitability of their growth and glory, until they, too, come crashing down. Historically, imperial ambitions have been kept in check by competing empires or by resource limits.

History is littered with emperors who were absolutely corrupted by their absolute power, and who brought their empires crashing down about them. Whether the House

---

[6] Founded in 1952, Edelman is a global communication firm that partners with businesses and organizations to evolve, promote, and protect their brands and reputations. "Edelman's trust research, the Edelman Trust Barometer, turns the deep data we collect into real-world insights; our trust consulting platform, Edelman Trust Management, interprets those insights to help our clients plan, make decisions, and take action; and our research institute and learning laboratory, Edelman Trust Institute, publishes data-driven insights that inform leadership, strategy, policy, and sustained action across institutions." (https://www.edelman.com/trust/2023/trust-barometer).

of Hapsburg that ruled the Austro-Hungarian empire and came crashing down in 1918, or the final Russian czar Nikolai II Alexandrovich Romanov who was deposed by a revolution in 1917, the Axis powers of Italy, Germany, and Japan who were defeated in 1945, or the Soviet Empire in 1989, empires driven by unfettered growth *always* overstep their bounds. Many other examples from outside of Europe and North America, whether the Tokugawa emperors of Japan or the Qing dynasty of China, who extended their rule into Manchuria and the Korean peninsula until they came crashing down with the Boxer rebellion. Today territorial conquest is mostly a relic of the nineteenth and twentieth centuries (and before), to be replaced by technological empires. Russia's missteps in Ukraine make this clear: off-the-shelf drones and sophisticated cyber weapons have delivered significant damage to tanks and rifles and become decisive tools.

Technological advances, whether of transportation (automobiles, railways) or production (factories and supply chains) or media (radio, television, and now the Internet), have *always* created regulatory challenges. As railways knit together the continental United States, the Interstate Commerce Commission, created in 1877, to regulate *common* carriers, later extended to trucking lines. With the popularization of the "horseless carriage" in the early twentieth century, challenges were created for urban planning which the community studies of the emerging Chicago School (see above) outlined. With the growth of broadcast media in the mid-century, the Federal Communications Commission created in 1934 as radios proliferated in households. As discussed in a previous chapter, Congress created the Fairness Doctrine of 1949 requiring broadcasters to give equal time to opposing points of view. In the previous century, the knitting of the continent through transportation technologies required new forms of regulation; in the current century the knitting together of the world requires not simply new laws but an entire new imagination of international regulation and in fact a reimagination of the nation-state. Calling these "regulatory challenges" minimizes their challenges to society. These might be better understood as *existential* challenges that question "who are we?" Are we resigned to being slaves of our own devices, or do we have the intellectual and moral and political resources to step back into the saddle?

To reclaim the common good in a technological society, we need to reawaken from our collective enchantment with technology and find the political will to hold private entities accountable for the damaging consequences of hegemonic technology and devote our attention and energies to rebuilding civility in the world we have lost.

References

Armstrong, A. K., Krasny, M. E., & Schuldt, J. P. (2018). *Communicating climate change: A guide for educators*. Cornell University Press.

Benjamin, R. (2019). *Race after technology: Abolitionist tools for the new Jim code*. Polity Press.

Chomsky, N. (2009). The mysteries of nature: How deeply hidden? *Journal of Philosophy., 106*, 167–200. https://doi.org/10.5840/Jphil2009106416

Davis, A., Gardner, B., & Gardner, M. R. (1941). *Deep south: A social anthropological study of caste and class*. University of South Carolina Press.

Edelman Trust Institute. (2023). *2023 Edelman Trust Barometer*.

Ellul, J. (1964). *Technological Society*. A. A. Knopf. (Original work published 1954).

Fromm, E. (1941). *Escape from freedom*. Farrar & Rinehart.

Gell, A. (1988). Technology and magic. *Anthropology Today, 4*(2), 6–9.

Hayek, F. (1944). *The road to serfdom*. University of Chicago Press.

Hess, D. (1995). *Science and technology in a multicultural world: The cultural politics of facts and artefacts*. Columbia University Press.

Kunstler, H. (2012). *Too much magic: Wishful thinking, technology, and the fate of the nation*. Atlantic Monthly Press.

Lynd, R. S., & Lynd, H. M. (1929). *Middletown: A study in contemporary American culture*. Harcourt.

McKibben, B. (2008). *Deep economy: The wealth of communities and the durable future*. One World Publications.

Meadows, D. (1972). The limits of growth: *A report for the club of Rome's project on the predicament of mankind*. Universe Publications.

Packard, V. (1959). *The hidden persuaders*. Pocket Books.

Park, R. E., & Burgess, E. (1925). *The city: Suggestions for the investigation of human behavior in the urban environment*. Springer.

Reisman, D., Glazer, N., & Denney, R. (1950). *The lonely crowd: A study of the changing American character*. Yale University Press.

Rose, D. (2015). *Enchanted objects: Innovation, design, and the future of technology*. Scribner.

Seeman, M. (1975). Alienation studies. *Annual Review of Sociology, 1*(1), 91–123. https://doi.org/10.1146/annurev.so.01.080175.000515

Stein, M. (2015). The eclipse of community: An interpretation of American studies. *Princeton University Press*. https://doi.org/10.1515/9781400868476

Stuckey, R. (2013b). *Cyber bullying*. Crabtree Publishing Company.

Stuckey, S. (2013). Slave culture nationalist theory and the foundations of black America. Oxford University Press. pp. IX, XI, XX, 44, 106–107.

Turkle, S. (1995). *Life on the screen: Identity in the age of the internet*. New York. Simon and Schuster.

Warner, W. L. L., J. O., Lunt, P. S. S., Leo. (1963). *Yankee city*. Yale University Press.

Chapter 14
Where We Go from Here

Abstract In this chapter, we examine how a new society can be constructed beyond the technological hegemony that we have lived with for nearly a century. We examine how the breakdown of society in terms of trust and shared narratives (such as science) and the rise of vulnerability, uncertainty, complexity, and ambiguity potentially lead to a societal collapse. The emergence of Artificial Intelligence and rising questioning of technological hegemony, as exemplified by the journalism of Kara Swisher, understanding the *context* of technology is no less important than understanding the devices themselves. The *choices* that we make with regard to our technological objects are no less important than the form and function of the objects themselves.

The title of this chapter suggests a "turn" that redirects technology away from its current hegemonic characteristics and trajectory. What we offer in our conclusion is not an answer but a sketch of an emerging paradigm of *what might be* based on the weak (and strong) signals[1] in contemporary discourse. Our goal was to lay the groundwork to show how the power and impact of today's technology have created technological hegemony. We have shown it is not the "thing" (technology) itself but the forces and conditions that have crafted the narratives, facilitated the choices, and enabled human and nonhuman actors to create our relationships with technology today. We pose another question: *Who are we* in our relationship with technology?

In Chapter 8, we asked, "Who are we?" and explored the way in which we tell the stories that propose to answer this question. Crafting narratives, whether our individual identities, our families, or for a nation is a collective venture. A *shared* story can fragment into many different stories as factions within society attempt to control the narrative and tell the story of *their* version of reality. Fact is not required in this process of constructing reality when there is no consensus as to what the facts are.

[1] The concept of weak signals was introduced by Igor Ansoff in his book *Strategic Management* (1975). In *Sensemaking in Organizations* (1995), Karl Weick expanded on the idea within the context of organizational sensemaking. Weick's research focuses on how organizations interpret and respond to weak signals, emphasizing the processes of sensemaking and the importance of noticing and interpreting weak signals to anticipate and adapt to changes in the environment.

A. Batteau and C. Z. Miller, *Tools, Totems, and Totalities*,
https://doi.org/10.1007/978-981-97-8708-1_14

We are living in a society of competing technological paradigms where the dynamics of power is the determining factor. Power is not shared equally, and choice is often not an option. “Where we go from here” is a story that is still being written. Each question about the trajectory has endless answers each posing more questions depending on an endless array of variables. Most likely, a clearly defined path forward was never possible but not so long ago—relatively speaking—there was a sense that “the future” would fundamentally resemble the past. Following a traditional story arc, the present would unfold at a somewhat manageable pace that would allow us (humans) to adapt. If this was ever the case, it certainly is not so today. We are facing the situation that many species that cohabit this planet have faced in the past and are facing today: adapt to rapidly changing environmental conditions *or else*.

Weak and Strong Signals: Rapid Technological Development

To put things in context, consider the grand challenges/major threats confronting the world today. In its 2024 Global Risk Report,[2] the World Economic Forum (WEF) lists “a deteriorating global outlook” as a key finding. Saadia Zahidi, WEF’s Managing Director, noted.

> Last year’s Global Risks Report warned of a world that would not easily rebound from continued shocks. As 2024 begins, the 19th edition of the report is set against a backdrop of rapidly accelerating technological change and economic uncertainty, as the world is plagued by a duo of dangerous crises: climate and conflict. (2024, p. 4)

Based on extensive research and survey data, the 2024 WEF report lists *misinformation and disinformation* as the most severe global risk[3] over the next 2 years with *extreme weather events* a close second. If we do not seriously address these two, the others—social polarization, cyber insecurity, interstate armed conflict, famine, mass migrations—will be exacerbated. And this is assuming humanity has a modicum of control over other events such as acts of Nature or God.

The 2024 Edelman Trust Barometer[4] is released annually by Edelman, a public relations consultancy and global relations firm founded in 1952. Published in the same timeframe (January 14, 2024) as the WEF Global Risks Report, the 2024 Trust Barometer reports “A collision of trust, innovation, and politics.” Along with a decline in trust in companies based in global powers, respondents to the survey “see science as under political pressure, but feel government lacks the competence to regulate innovation effectively, so strong leadership is needed to move society toward

---

[2] World Economic Forum (WEF) https://www3.weforum.org/docs/WEF_The_Global_Risks_Report_2024.pdf

[3] Figure C: Global risks ranked by severity over the short and long term (2024, p. 8).

[4] Edelman is a global communication firm that works with businesses and organizations to evolve, promote, and protect their reputations. https://www.africa.edelman.com/expertise accessed July 17, 2024.

acceptance." This year's report suggests a correlation between rapid technological advances is outpacing people's capacity to adapt to disruptive technological change.

> The 2024 Edelman Trust Barometer reveals a new paradox at the heart of society. Rapid innovation offers the promise of a new era of prosperity, but instead risks exacerbating trust issues, leading to further societal instability and political polarization.
>
> In a year where half the global population can vote in new leaders, the acceptance of innovation is essential to the success of our society. While people agree that scientists are essential to the acceptance of innovation, many are concerned that politics has too much influence on science. This perception is contributing to the decline of trust in the institutions responsible for steering us through change and towards a more prosperous future.

The rapid pace and disruptive nature of innovation is one factor in the increasing fragmentation of society. The ability to adapt to disruptive technological change on the individual, corporate, and government levels is destabilizing. VUCA—vulnerability, uncertainty, complexity, and ambiguity—is a concept adopted by the U.S. Army War College (Barber, 1992; Bennis, and Nanus, 1986) to describe the strategic environment confronting military leaders. Businesses and organizations have since adopted VUCA to describe the environments they operate.

Disruption and VUCA conditions can exit at the societal level as well as the global level. Laureston Sharpe's (1987) ethnographic essay "Steel Axes for Stone Age Australians" provides insight as to why the introduction of a steel age artifact (an axe) into a stone age society resulted in profound repercussions. Not only was the stone axe an essential tool for the Yir Yoront, it was also a central component of their patriarchal social hierarchy. The steel axe was introduced by Western missionaries in efforts to "win over" the indigenous population. Steel axes were especially attractive to women and young men who were not allowed by Yir Yoront societal norms and practices to make and possess the stone axe. The allure of greater independence and agency along with perceived functionality of the steel axe was a hit. However, displacing the stone axe, a key cultural artifact in Yir Yoront society and totemic system, and intricately tied to the stories of "Who we are", with a Western style steel axe led to societal collapse over the space of just a few generations. As a society, the Yir Yoront were unable to adapt to the forced march from stone age to steel age precipitated by the introduction of too much innovation, too fast.

Even today, it is not uncommon to hear traditional indigenous societies referred to as "primitive" as opposed to our modern societies. Yet the more we learn about these so-called primitive societies, ironically often aided by advanced technology, the more we are confronted with how much we have in common as humans, how much we *are* them. Contemporary (modern) societies can be disrupted by an accelerated rate technological change. The challenges societies face in adapting to rapid technological change, and the effect on human physiology is a topic of debate. Can we keep up? Do we want to?

The disruptive change shaking the foundations of society's long-standing institutions is not the result of any one thing, but rather a series of events that have triggered a cascade of consequences, each igniting new cascades of events that resonate on multiple levels. Unlike other species,[5] humans have played an outsized role in creating the environmental conditions we currently face in the unofficial but aptly named "Anthropocene" epoch.[6] For better and for worse technology has been a major actor as well as our intimate companion, ever extending our physical and cognitive abilities and shaping the natural and artificial worlds we inhabit.

Writing Technology

The heading for this section is a reference to *Writing Culture: The Poetics and Politics of Ethnography* a collection of essays edited by James Clifford and George E. Marcus (1986/2010) originally published in 1986. The discussions held at the School of American Research in Santa Fe, NM in April 1984, looked critically at "one of the principle things ethnographers *do*—that is write…" We carry on the principle activity of *writing* in this ethnographic exploration of technology, looking critically at the ways in which technology has evolved, and how it has been shaped by institutional and social frameworks. We also consider how the growing dominance of technology has put the *value* of this essential ethnographic practice of writing in question. What is the value of writing when *reading* is no longer an activity that has value? This cuts to the heart of concerns regarding anthropology's relevance today. Many anthropologists—students, newly minted and legacy practitioner/scholars—are deeply engaged in reinventing how to communicate the value of anthropology in a world *beyond text*, the traditional vehicle of ethnography.

Writing as we did in the aftermath of COVID and the ongoing post-pandemic hangover, we would be remiss to conclude without acknowledging two major technology-related events that occurred as we were working on this book. The first event, of course, was the emergence of Artificial Intelligence or AI. In June 2020,[7] the American Artificial Intelligence Research Organization, better known as Open AI, launched Chat GPT, a large language model able to achieve general purpose language generation and natural language processing tasks. Funded in large part by Microsoft, many other companies immediately followed suit to develop their own LLM (large language models) launching an AI gold rush. Some, like semiconductor

[5] Unless the species in question is out of balance in their environment, for example in Yellowstone National Park, the extermination of wolves to protect deer and elk populations resulted in multiple negative consequences for the park environment and ecosystems. (Peterson, 2020).

[6] The current official unit of geologic time is the Holocene epoch that began after the last ice age roughly 11,700 years ago. The unofficial epoch referred to as the Anthropocene began around 1950 and "describes the most recent period in Earth's history when human activity started to have a significant impact on the planet's climate and ecosystems." (*National Geographic*, 2023).

[7] Date was checked by Chat GPT.

producer Nividia,[8] experienced exponential growth as business from all sectors of the economy rushed to embed AI into their business and organizational operations.

The second event was the publication of *Burn Book: A Tech Love Story* (2024) in February 2024 by technology journalist Kara Swisher proved to be another important marker.[9] Swisher's book is personal, documenting her coming of age in the age of increasingly powerful computers, the emergence of the Internet, and the rise of a class of entrepreneurial technology titans including Mark Zuckerberg, Steve Jobs, Elon Musk, Jeff Bezos, Jack Dorsey, Bill Gates, and many others. Her prescience is impressive: Swisher correctly anticipated the coming of "the digitization of everything." As a young technology journalist Swisher covered a beat that few saw as relevant. Her long partnership with Walt Mossberg, principal technology columnist for the Wall Street Journal (WSJ) from 1991 through 2013, Swisher not only occupied a front row seat, but was often *on* the stage with first movers in the nascent "high tech" industry. In fact, she *built* the stage along with Mossberg co-founding podcasts and conferences including *All things D*,[10] *Recode*,[11] and the D and the Code Conferences. Commenting on the meteoric rise of elite tech founders of a "scrappy upstart industry into leaders of some of America's largest and most influential businesses" Swisher notes:

> When people get really rich, they seem to attract legions of enablers who lick them up and down all day. May of these billionaires had then started to think of this fawning as reality, when suddenly everything that comes out of their mouth is golden. History gets rewritten as hagiography.[12] But if you know them in the before times and have some prior knowledge of their original lives, you either become an asset (truth-teller) or a threat (truth-teller) to them. (Swisher, 2024, p. 4–5)

Including Swisher's wit, down-to-earth language, and unflinching commentary brings dimension and additional threads to the complex layers of the conversations in this book.

---

[8] According to Companies Market Cap.com, "Nvidia Corporation is one of the largest developers of graphics processors and chipsets for personal computers and game consoles. The head office is in Santa Clara, California. As of July 2024 NVIDIA has a market cap **of $2.902 Trillion**. This makes NVIDIA the world's third most valuable company by market cap according to our data." https://companiesmarketcap.com/nvidia/marketcap/ accessed July 17, 2024.

[9] "Burn book" is a combination of two slang terms: burn (used as a verb), to insult, and book, a stand-in term for gossip and malicious behavior. The term was used by characters in the 2004 film **Mean Girls** to record and spread gossip about their classmates. https://thecinemaholic.com/mean-girls-ending-explained-what-is-the-burn-book/ accessed July 17, 2024.

[10] Co-founded by Swisher and Mossberg in 2007, an online publication that covered technology and tech startup news and analysis that later hosted *D: All Things Digital Conference.*

[11] From 2014 to 2019 technology news, website co-founded by Mossberg and Swisher that was focused on the business of Silicon Valley. Recode was purchased by the Vox media network.

[12] Hagiography is a form of writing that idealizes its subject such as the Catholic way of writing the lives of the saints.

Technology and the Modern Imagination

We began the Introduction with the provocative claim that "technology's contribution to human progress is mostly imaginary." We make this claim because the imagination that perceives an axiomatic equation between "technology" and "progress" stands *only* as long as it is not called into question. We do not dispute the wonders of technology or "modern technology". What we call into question is the notion that technology unequivocally represents an advancement in the human condition. We interrogate the belief that technology enables "a god-like transcendence of the mortal human condition" and argue that this belief must be seen through the lens of Western civilization.

It's Never Just One Thing

Although technology is the focus of this book, we have been vigilant to avoid presenting technology in isolation, as a *thing* without context. Our concerns include the consequences of the modernist paradigm of technology that equate technology with "progress". The distinction was made in the introduction between "useful things" such as eating utensils, hoes, rakes, and axes that were designed for specific purposes and "high technology", which carries the mystique of efficiency and "progress". And yet we recognize that even so-called primitive technology has embedded meaning that can be disrupted by changes in materials and design.[13]

We know that progress and the benefits of advanced technology are not universally distributed. Progress for whom? At what cost? Concepts such as "externalities" and "consequences" have cropped up in discussions of economics and are creeping into financial cost–benefit analysis. When compelled by either customers, stakeholders, or regulations to account for the true or actual costs beyond (i.e., external to) those of the enterprise itself the math looks very different. We are left with the uncomfortable question "Who will pay?".

As Kara Swisher noted in the first line of Burn Book, "It was capitalism after all." (2024: 1). However, as we have already argued, it is never just one thing.

[13] We have provided an example of the power of embedded meaning is highlighted in Laureston Sharpe's (1960) ethnography of the Yir Yoront people of the Cape York Peninsula in Australia. The introduction and diffusion of steel axes that displaced stone axes is a case study of how a society can be disrupted and destabilized by a "better" technology. To the change agents, innovators, and early adopters of "steel axes" this is a cautionary tale.

In Conclusion

Under conditions of relative stability organizations and institutions tend to change slowly. Depending on the availability or lack of resources, for an entire species to evolve can take hundreds of years. Or not. In fact, massive change can occur quickly. And violently. In the meantime, opportunities to choose another path go unheeded. Evolution can manifest as a physical trait, a mental ability, or an emotional response. Adaptation is always an experiment and can be anything from burrowing into the ground to sprouting wings or learning to swim. In any case, survival is not guaranteed.

As humans, we tend to fall in love with our technological inventions. We marvel at them as if they did not come from us, as if they were not *of* us. We also tire of them as the next new thing is presented with little or no regard for consequences. Our only thought is "what will it do for ME?" How will it serve MY ends?

It is possible to evolve a collective consciousness. We know it possible to shift from "me" to "we" because other cultures—and American culture—have gone through periods of high civic engagement, coming together for common purposes. Under the right conditions and circumstances, the answer is *yes*. As political scientist Robert Putnam explains, this is also part of the American story *who we are* (Putnam, 2020, 2021). In *Bowling Alone,* (Putnam 2000) describes the decline of civic engagement in the post-World War II years, as Americans became less community-oriented and more consumption-oriented. Twenty years later, in *The Upswing* (2021), Putnam presents a more positive comment, noting that Americans were rediscovering and rebuilding communities, investing in social capital beyond simply consumption; *social capital*, the investments that people have in their kin, friends, and neighbors, is a better foundation of a good society than simply piles of gold. History has shown how self-preservation and the idea of "The Greater Good" have been abducted in some cases to shape our behavior in a hateful, self-destructive fashion. Yet this is not the only story. In either case, it is always a choice.

We are faced with these choices today, both individually and collectively. Yet it is not only for us that we choose. Humans share the earth with other species but have an outsized amount of power to impact and shape the environment. We know this. With great power comes great responsibility.

More can be written, discussed, and reasonably debated, but better we consider how to collectively, ethically, and sustainably design futures in which people are not relegated to the status of "users", where technology is intentionally designed to allow for choice rather than conscription or a forced march, and it enhances the prospects of survival for all beings and the planet.

References

Ansoff, H. I. (1975). *Strategic management*. Palgrave Macmillan.
Anthropocene. (2023). *National geographic*. Retrieved March 19, 2024, from https://education.nationalgeographic.org/resource/anthropocene/

Barber, H. F. (1992). Developing strategic leadership: The US Army War College experience. *Journal of Management Development, 11*(6), 4–12. https://doi.org/10.1108/02621719210018208

Bennis, W. G., & Nanus, B. (1986). *Leaders: The strategies for taking charge.* Harper & Row.

Edelman. (2024). 2024 Edelman Trust Barometer. Retrieved July 17, 2024, from https://www.edelman.com/trust/2024/trust-barometer

Peterson, C. (2020). 25 years after returning to Yellowstone, wolves have helped stabilize the ecosystem. *National Geographic.* https://www.nationalgeographic.com/animals/article/yellowstone-wolves-reintroduction-helped-stabilize-ecosystem

Putnam, R. D. (2020). *Bowling alone: Revised and updated: The collapse and revival of American community.* Simon & Schuster.

Putnam, R. D. (2021). *The upswing: How America came together a century ago and how we can do it again.* Simon & Schuster.

Sharp, L. (1987). Steel axes for stone age Australians. In J. P. Spradley & D. W. McCurdy (Eds.), *Conformity and conflict: Readings in cultural anthropology* (6th ed., pp. 389–411). Little, Brown and Co.

Swisher, K. (2024). *Burn book: A tech love story.* Simon & Schuster.

Weick, K. E. (1995). Sensemaking in organizations. Sage Publications.

World Economic Forum. (2024). *The Global Risks Report* (19th ed.). WEF.

Index

A. Batteau and C. Z. Miller, *Tools, Totems, and Totalities*,
https://doi.org/10.1007/978-981-97-8708-1

Printed in the United States
by Baker & Taylor Publisher Services